THREE JOURNEYS

An Automythology

THREE JOURNEYS:

An Automythology

PAUL ZWEIG

Basic Books, Inc., Publishers

NEW YORK

Library of Congress Cataloging in Publication Data

Zweig, Paul.
  Three journeys.

  1. Zweig, Paul—Biography. I. Title.
PS3576.W4Z52   818'.5'409 [B]   75-36386
ISBN 0-465-08610-1

Copyright © 1976 by Paul Zweig
Printed in the United States of America
DESIGNED BY VINCENT TORRE
76 77 78 79 80   10 9 8 7 6 5 4 3 2 1

To Baba

CONTENTS

Preface

ix

PART ONE

Against Emptiness

1

PART TWO

Automythology

63

PART THREE

The Bright Yellow Circus

139

PREFACE

IN the spring of 1974, I spent a month travelling alone in the Sahara Desert. Planning for the trip had begun casually with some friends, a year before. But later my companions began to have second thoughts, and I was surprised to discover in myself a determination to go anyway, even alone. Travelling to the desert had suddenly become extraordinarily important to me.

I devoured geology text books, North African history books, all the travel stories I could get my hands on. I spent hours staring at maps, wondering at the large blank patches where nothing worth printing was apparently known about land features. I could find out little about actual travelling conditions. Different writers, referring to the same stretch of track, would describe it as dangerous or as an easy promenade. People I spoke to had opposing opinions on virtually everything. Mostly they wanted to talk about snakes and scorpions, and I began to think of the desert as a vast crawling place.

My determination to go, even if no one else could, seemed strange and a little grim, even to myself. I was dimly aware that my age had something to do with it. I was thirty-eight, and had gotten much of what I'd spent a great deal of time wanting. I loved my wife; I had a reasonable job; several of my books had been published and praised. All this was good, but somehow it wasn't good enough, or maybe I wasn't good enough. I suspected that I didn't know how to want properly, that I had a way of giving things back merely by having them.

To make this journey would be to seize hold of possibility. I would discover at the last minute that I was another kind of person

Preface

after all; for example, one who could go travelling in the Sahara Desert. In a way, these were not thoughts at all, but attempts to make my determination respectable by finding reasons for it. When I scratched a little, I had to admit that I didn't know why I wanted to go.

In the end I went alone, and the aloneness became the trip. The journal I kept, composing the first of my "journeys," was part of it. I wrote in the journal almost every day, not so much to keep a record of my experience as to preserve a life line, a sort of mental tent I could withdraw to now and then, in the stunning simplicity of the desert spaces.

When I found myself once more amidst streets, people, trees, needs, I also found waiting for me the question I hadn't answered: why had I gone into the Sahara? Why had the desert become for me a luminous personal space? Of what inner story had it become the climax, and what next?

A few months after returning from the desert, while writing about it, discovering that the answer to my question was, in some sense, all of my life, in particular the decade I had spent living in Paris—not sure at that point if I was writing a book, or an extended letter to myself—I met an Indian Sadguru, a holy man, Swami Muktananda. Despite various political adventures, I had always thought of myself as a cultural conservative, a resolute, even a militant, Westerner, with a passionate stake in Western cultural history. I had previously taken no more than a sociological interest in the "Eastern religion scene," deploring what I felt to be a peculiarly American weakness for spiritual melodrama. I visited Swami Muktananda on the urging of a friend, and was overwhelmed, not so much by the man—I didn't begin to grasp his nature until later—as by my own puzzling emotions which triggered a series of intense private experiences forming the third of my "journeys."

PART ONE

Against Emptiness

April 3

I APPROACHED THE SAHARA this morning across a landscape of metallic rocks and gnarled half-trees. The occasional peeled patches of vegetation seemed almost repulsive, and I felt relieved when they disappeared, replaced by coppery sand flats and endless fields of pebbles. The road as far as Bechar runs along a fortified border built by the French during the Algerian war for independence. On both sides of the road tangles of barbed wire unravel for hundreds of miles, punctuated by mounds of broken green glass and empty blockhouses. The French moved the Berber population out of the area during the war, and few seem to have come back. The railroad stations and clumps of buildings around them are in ruins, despite the eery persistence of signposts giving freshly painted names to the destroyed settlements.

After leaving Bechar I drove for a hundred miles across a gravel plain the color of gunmetal. Its dark absorbent surface gave the light a toned-down almost depressed quality. A few hours later I topped a hill and saw, facing me across a narrow valley, a wall of rust-colored sand which seemed to radiate light from within itself. At the foot of it, the tiny oasis of Taghit.

I'm camped about a mile from the *ksar*, on the edge of the palm grove. The desert silence is crisp and stonelike. The almost full moon sheds a bluish light, into which saw-edged palm leaves are cut. Even now the dune gives off a hint of its inner rust. Small shadows, cast by flat stones balanced on one end, speckle the ground in front of my tent, marking the graves of an Arab cemetery. The haphazard scattering of graves contrasts with the smooth outline of the dune rising above the plain a few hundred yards away.

April 4

WE'VE GOTTEN THROUGH the first day—the car and I—at the cost of exhaustion and a layer of reddish yellow dust over everything. Shortly before dusk I set up my tent in a bowl of crumbly brown gravel between the oases of Igli and Beni Abbès, after inching for a hundred miles over an excruciating track. Night has fallen in minutes, and the low ridges of the basin resemble an irregular black wave against the hazy remnant of daylight. There is no wind at all, no sound, except for the scratching of my pen. A while ago I heard a faint creaking noise in the distance and looked up as a wedge of birds flew by, their lonely *V* of movement almost invisible against the darkening sky.

The shopkeeper in Taghit had warned me not to take the track to Igli. He said it was deeply rutted and littered with sharp flinty rocks. Besides, no one used it any more. If I broke down, I might have to walk as much as fifty miles. I started late in the morning, intending to make a 250-mile detour by road. But it depressed me to pass the cairn of dull black stones marking the track.

I knew before I started that travelling alone would require special caution. For example, I would have to keep away from badly marked secondary tracks, where no one might come by for days if I were hurt or had car trouble. But it was frustrating to see that ribbon of creamy dust disappear behind a hill. Why didn't I try it for a few miles, and turn back if the going became dangerous? It occurred to me as I swerved onto the ripply washboard surface that I might not know what danger looked like if I saw it. The track was thick with dust and horribly uneven, but it was clearly marked in most places, and there didn't seem to be a great deal of sand. Although the car rocked like an earthquake, it held its own without too much trouble. The trick is apparently to go slowly, as little as five or ten miles per hour.

Occasionally the track swung close to the yellow crests of the Erg. The contrast between the irruption of black fragments I drove through, and the smooth peacefulness of the sand sea, was startling.

April 5

I'VE BEEN THINKING about some of the people I've come across during the past few days. On the track to Igli I passed a man in a tattered white jellaba, leading a camel. He motioned me to stop and, to my disappointment concerning the purity of desert customs, offered to let me take his picture for a cigarette. He too was headed for Igli, at least a four-day walk for him, maybe more. His water supply was carried in a patched inner tube wound around his camel's neck and saddle. It must have contained several gallons. When I left, he shook my hand with a solemn simple motion, then touched his chest and lips with his right hand, in the Moslem gesture of greeting.

A human presence in the desert is so unexpected, it is almost dreamlike. The smooth dark rocks and wide reaches of gravel seem to leave no place for the vulnerability of human life. So it was ghostly to make out the far off miniature of a man leading a camel. Yet when I stopped the car and we smiled and spoke sign language to each other, the desert lapsed, that is the only way I can put it. It became smaller, less razory. And it clamped over me again, like a bright bowl, the minute I drove off.

Last evening, as I was making camp, a white-robed figure came trotting toward me out of the gravel basin. I had noticed several camels grazing on scattered tufts of bush in the distance, and he, clearly, was their keeper: a young boy, maybe ten or twelve years old. He made the Moslem gesture of greeting, and set to work without a word, helping me to mount my tent. All his movements were discreet and gentle. He made a sign asking for water and thanked me with a relieved happy smile which lit up his entire face. He accepted the cigarette I offered him with the same expression of surprised pleasure and put it carefully away in the leather pouch he carried about his neck. A while later I watched him herd his camels together and drive them between two hills. Clambering up a slope, I saw in a neighboring valley other small camel herds moving at an immense crawl in the same direction, toward a place several miles off where the nomad camp was located.

A little while ago, a car rumbled by under a trail of dust about half a mile away. It disappeared into a fold in the ground, and then came circling back toward my tent. The sight of a dwarfed gray car chugging over a plain full of light was almost comical. When it reached me it stopped, and four men got out carrying some glasses and a bottle of Ricard. They turned out to be the schoolteachers and town clerks from Igli, the oasis I passed through toward the end of the day yesterday. They had seen me go by, and supposed I must have camped somewhere in the area, so they drove out this morning to have a drink with me. I asked them if drinking conflicted with their Islamic precepts, and they smiled a little shyly, saying it was a small sin, but one they appreciated. I brought out some crackers, and we sat on the slowly warming sand in front of the tent satisfying as much of our mutual curiosity as we could manage. I asked them about the Reguibat nomads who had been seen in the dunes near Taghit and Igli. "All the Reguibat are interested in is money," they spat out scornfully, and laughed. They were guarded when talking about Algerian politics, but one teacher surprised me by saying, "To tell you the truth, we envy the Moroccans around here." I told him I thought Morocco was a corrupt, unhappy country. I didn't understand what there was to envy. "At least they have a king," he answered, only half jokingly, and I thought I understood the poignancy of his joke. Whatever personal qualities a king might have, he provided a bond between the larger world and the lives of ordinary people. He represented a horizon, a dream (if only a daydream) of grandeur. Even if the king was corrupt and incapable, at least he was a living person, unpredictable, extravagant. In Algeria there are no kings, no heroes. The needs of social justice are better served, and the poor (which is practically everyone) can be thankful for real benefits. Yet the system makes it harder for them to expand beyond their circumscribed existences, except maybe in the practice of religion. And it is true, in the desert Islam is omnipresent. When I passed through Igli at dusk yesterday, the shadowed edge of the main street was lined with men in jellabas kneeling face down to greet the night advancing toward them from Mecca, in the east. But even in the desert, religion is a waning influence among the more comfortable classes. In a way, it seems better to be poor in Algeria, for the poor still possess a ceremonial framework. But these teachers and and clerks were lost men. Their dilemma was sharpened for me by something else they said: "We're divided, you see. Our hearts are with our brothers in the east, yet we prefer to talk to you from America, and we listen every night to the French radio." They were suspended between nowhere and nowhere, holding on to the only ritual they had left: the hospitality which they performed with such grace. I found myself feeling great tenderness for these men sitting in the gradually warming emptiness of early morning.

They drove off a few minutes ago, leaving behind their footprints, like an intricate mandala cradled by the wide gravel surface of the valley.

April 6

I'M CAMPED on the edge of a cliff overlooking the *oued* Saoura. The wind has been puffy all day, but now it has died away. While I sip my tea, my legs dangling over the glowing cleft of the *oued,* the light has suddenly dampened and changed color. The sun lies far upstream between the cliffs. Everything has become orange: the dunes marking the edge of the Erg, the gravel and dried mud interspersed below me in the *oued*. Even the air seems to have turned orange.

April 7

I'VE NEVER SEEN LAND as bleak as what I drove through yesterday. Two hundred miles without a trace of life: no palm trees, no scrub, no people. After Kerzaz the road wound through a maze of stark black cliffs, the level portion of the valleys strewn with jumbles of rock, or else, in smoother stretches, with woundlike swellings puffed up into a hideous mimicry of softness. After fifty miles or so, the road drew closer to the Erg, and the folds of rock became streaked with sand.

In more temperate climates, vegetation, erosion, and man have worked out a sort of harmony which we call natural. It's hard to think of a forest or meadow being in bad taste. But in the Sahara, nature still possesses an archaic prehuman ugliness: rusty dunes, beds of dried blinding mud, rocky waste heaps strewn for miles, interrupted at huge intervals by specks of watered soil which are the oases.

That this mineral profusion has allowed patches of vegetable and human life to develop at all seems to be a mystifying trick. When you jolt over a hill and see a dark streak at the base of the next hill, dusty green interspersed with flashes of geometrical red wall, you are afraid to look away, because by every law of probability it shouldn't be there, and may not be the next time you look.

I don't imagine even island peoples are as isolated as the population of the oases, at least before motor vehicles and roads were introduced, less than forty years ago. To cross the hundreds of miles from one oasis to another required expert skill, along with a dose of madness. The margin for error was measured by the amount of water a camel could carry, extended slightly in an emergency by the camel's own blood. In summer, with humidity close to zero and the ground almost hot enough to cook meat, there was no margin. Before the camel was introduced to the western desert, probably during the late Roman Empire, communication between oases must have been rare, making use of oxen, perhaps horses, both of which are far more vulnerable to the extremity of desert climate than the camel. *Oasiens* must have believed themselves to be the only humans on earth. They must have believed that life was an insecure condition which could endure in no form at all without a sort of medical nurturing. The struggle was not between domesticated life and wild life, gardens and forests, it was between life itself, and the mineral chaos which stared at it from every side.

This explains the prestige of the nomad tribes which ruled the Sahara until not long ago. The nomads had mastered a unique skill enabling them to move around in the desert as boats move around in the sea. It was a skill composed of incredible physical endurance developed during centuries of adaptation to an awesome environment, and a willingness to drive the organism to extremes of suffering. It depended, therefore, upon a culture of fierce, unrelenting willfulness. It depended also on the monopoly of camels, which fell to them by default. To own camels, one must be able to feed them. To feed them, one must find pasture. But pasture in the Sahara is almost nonexistent. To an unaccustomed eye, desert pasture seems no less barren than any other stretch of rock and sand. It takes many square miles of desert to feed one camel. The Reguibat nomads in the dunes near Taghit had come 500 miles from Mauretania because they heard rain had fallen there.

These qualities enabled the nomad to cast his network of customs and needs across the entire expanse of the desert. The *oasiens* became his slaves not only because they were afraid of the nomad's warrior skill, but because the nomad monopolized another, more than human skill: the skill of movement, of which the greatest single expression was the migration of the Beni Hillel tribe out of Arabia in the eleventh century. By the time it attained its greatest westward thrust, it had encompassed the Sahara in a thinly spun net, the twin strands of which were the Arab language and the genius, learned in Arabia's "empty quarter," of restless movement across hostile spaces.

The oasis of Timimoun is on a height overlooking a dried lake bed, or *sebha*, and beyond it, the dunes of the Western Erg, which is now to the north of me. The palm grove spills down the slope into the *sebha* along quite a distance. Water must be plentiful, because the garden plots are flourishing, and the grove is huge. The water is brought by *foggaras*—primitive tunnels which may be many miles long—but some may also be pumped from wells. The water flows in *seguias*—artificial channels of varying depth and width—divided progressively by the teeth of "combs" into ever smaller channels. The smallest channel empties into a single garden plot, where the water is stored in an open basin. The loss from evaporation must be enormous. The oasis crops are wheat, onions, tomatoes, and other garden vegetables; some cotton; and of course, dates.

The neighborhoods of the *ksar* seem to be roughly divided by occupation: sharecroppers (*khamès*), working the gardens and palm groves, live in the old *ksar* overlooking the *sebha;* camel-owners and merchants live in what is apparently a newer section, with wide rectangular streets opening into empty desert. Although slavery was abolished by the French half a century ago, no one seems to know whether it still exists or not. The social structure of Timimoun remains obscure, in part because so much of life here is leveled by misery.

April 8

TODAY I VISITED several small oases in the Gourara near Timimoun. The track was almost impassible in places, and I spent half the day digging myself out. The landscape seemed almost

Against Emptiness

Biblical: palm groves every few miles huddled behind mud walls to keep out the drifting sand; dazzling red geometries of granaries and houses piled beside them like crude building blocks. Along the track, white-turbaned men drove donkeys loaded with sacks of grain. Their glistening black faces stood out against the faded colors of the desert. Each *ksar* was imbedded in a maze of ruins trailing into the ground. When, every century or so, a cloudburst melts away the mud-built houses, the *ksar* migrates to an adjacent open space, adding itself onto the dilapidated puzzle of rooms and alleys which are converted into granaries, stables, or into wind-harps, inhabited by sand.

One *ksar* I visited was built against a low cliff with a spacious cave at the bottom. I was told that the villagers lived in the cave during the summer, when temperatures rose to 120 and 130 degrees. Centuries of restless sleeping had worn dozens of man-sized hollows in the sandstone shelving of the cave.

Toward the end of the day, returning across the *sebha*, I encountered a terrible sign of Saharan misery. Several miles before Timimoun, I had been told, the powdery sand contained a deposit of *roses des sables*—pink flowerlike crystals of gypsum, which can be quite large and beautiful. I stopped the car at a likely place, and began to poke at the ground. The crystals were not hard to find, and I soon had a pile of them, when I noticed, at a distance, a group of men huddled together on the sand. I thought of some dreadful enchantment in *The Thousand and One Nights:* a band of men condemned to survive alone in the arid mournfulness of the desert. I walked toward them and soon was able to make them out more clearly. Arranged on the sand before each man lay a barbarous pink frieze of *roses des sables*. The men hardly stirred as I came up, though I was probably the first person they had seen that day. The enchantment was of listlessness, indifference, hunger. They had beachcombed the desert for these magnificent crystals, and now were selling them for a few cents each. There must have been a dozen men squatting in the sun, and it was hard to believe that even one customer came by every day. Clearly a man couldn't hope to sell more than one or two pieces in a week, if that. I measured the abyss of their poverty. These men were adults, not children adding a few pennies to the family income. And they were willing to squat week after week in the shadowless sun, for a trickle of coins. Not enough to live on, barely enough to prolong their chronic illnesses, their listlessness, and their fatalism.

It is strange that Timimoun should be so beautiful, its red mud shaped lovingly by these dozing, undernourished lives into walls, alleys, roofs, pillars, gateways, bleach-white shrines, palm groves, intricate tunnels, irrigation combs, the passion-colors of gardens, succulent onion flowers, cotton, improbable green knobs of lettuce.

April 9

I'VE JUST SPENT the most exhausting day of my life. Until this morning the Sahara had been a gentle place: cold nights, crystal cool mornings; some wind, some heat. But today another Sahara showed itself to me; showed its claws, I wanted to say, but the image isn't right. One would have to invent some new animal faculty to describe the kind of assault I experienced.

When I woke up in Timimoun the sky was livid and a gusty wind was blowing. The air was strangely hot, even at six in the morning. By the time I had driven a dozen miles toward Adrar, the gusts had risen into a gale blowing out of the southwest, with a sound between a roar and a hiss. The wind kept getting hotter and stronger, sweeping along with it a layer of dust and sand which thickened until I could hardly see the road. I had to close the vents of the car and pour water over my head and clothes to keep cool. Here and there small dunes had drifted across the road and the car rattled menacingly when I drove over them.

It probably would have been wiser to pull off the road, turn the car back to the wind, and wait it out. But the land around was relentlessly flat, and there was no likelihood of it stopping before nightfall, maybe not even then.

After a few hours I made out a rectangular shape looming in the distance. The blur of dust and sand made it hard to see. The shape turned out to be a building with an improbable sign nailed to it: café. I drove the car behind the building, and went inside. It was a rough little room with a couple of shuddering tables, and a floor of sand. The scene was incredible. My eyes burned from the strain of the road, and I was covered from head to foot with powdery dust. A man who turned out to be the owner was sitting in the middle of the room on a stool, sawing erratically at an old violin. In a corner another man, in work clothes, was bent over a set of watercolors and a meticulous representation of a *ksar* in "primitive" style. For a bizarre minute, in the middle of the Sahara desert, scoured by a sandstorm, I thought I might be standing in a scene out of Eugene O'Neill. I collapsed onto a shaky bench, while the man on the stool sawed sporadically at his violin and smiled to himself with the timeless drowsiness of a *kif* smoker. The painter looked up and greeted me with elaborate Arab courtesy. He was from Oran in the north, he said, and had come to the Sahara to work on a road building crew, but now he was out of a job. He slept in the courtyard of the café, spent most of his time drinking tea and doing watercolors on a piece of white cloth. His painting was filled with camels, Douanier Rousseau palm trees, and the voodoolike figures of *haratins*—oasis slaves—with glistening black faces and arc-shaped toothy smiles.

After a few minutes the violinist stopped his squeaky playing and brought me a glass of bitter tea. He didn't speak any French, so we smiled back and forth at each other for a while. The walls of the building were so thick that one heard the wind only as a distant hiss. I drank glasses of tea, one after another, occasionally dozing off for a few minutes at a time. After about an hour, I gathered my courage and went outside to wash the air filter of the car in gasoline. At times the sand reached blizzard proportions, swirling and gusting around the edges of the low building. I came back caked with grit, drank some more bitter tea, and made patches of conversation with the painter. At last, fooled by a lull in the storm, I started out again, only to face the same choking wind, which did not let go until Adrar. It is still blowing. It may blow for days. The town is red, dusty and flat. About right for the end of the world.

April 10

LAST NIGHT was horrible. After the inch-by-inch panic of the road, a tornado of dust thicker than fog blew up in the middle of the night. The entire town rattled and banged and hissed. The streetlights in the distance were choked into a dirty brown clot. I threw some clothes on and ran down to the car to put a tarpaulin around the motor. I also tied plastic bags over the various locks, so they wouldn't choke up with dust.

Today the wind is still blowing, but it's cooler, not quite strong enough to raise the blanket of dust which made yesterday a bad dream. I'm told this is the season for wind. Today's muscular breeze could blow at any minute into the hellish wind of yesterday.

Yet as soon as the lid of dust lifts, the desert becomes beautiful again. I'm eating my lunch under an isolated wild palm near an oasis in the Touat region, between Adrar and Reggane. The *reg* has a gentle aspect, because of the scattered wild palm trees, and the dark green horizon of the palm grove a mile away. In front of me, in the distance, is the *ksar*, a zigzag of red mud walls with the inevitable white arch of a maraboutic shrine nearby.

The Touat is a ribbon of palm groves 80 miles long and no more than several hundred yards wide at its widest. It exists because an abrupt change in ground level brings the water table close enough to the surface to be reached by *foggaras*. All along the track one sees conical heaps of dried mud strung out in lines, marking the underground tunnels which supply water to the oases.

Against Emptiness

My camp is pitched on a depressing gravel plain which resembles my exhaustion. I can still feel yesterday's storm in my muscles. The track from Adrar started well enough this morning, but after a few miles it turned into one long ribbon of washboard. At first I tried to drive aggressively, at fifty miles an hour, but my nerves gave out. The car lurched crazily, and it became apparent I was going to break something, probably me. It turned out to be a tire which gave out a few miles later. After that, I resolved to take the coward's way out, and inch along peacefully at five to ten miles an hour. This way, at least, I will get the nomad's measure of desert travelling.

I am finding it difficult to keep up with my journal these days. The desert landscapes may be empty and still, but my experience is quite the opposite. It is intense, obsessive. I find it hard to remember even the previous day or hour, because of the grinding minute-by-minute need to pay attention. When, in the evening or at lunch, I want to write, I've got to remove myself from the intensity. I've got to think and remember despite the desert, which wants to absorb me in the sun sinking onto a plain of black pebbles, or in the daylong suspense as I lurch over a crusty clay surface, expecting it to give way and trap the car in the sand under it yet again.

The butane lamp sizzles as I lean over my notebook, trying to concentrate on the tip of my ballpoint pen weaving its familiar frieze of cramped letters, while the sides of my tent flap in the wind, making me remember that I am a hundred miles, maybe more, from the nearest human being; that my presence on this flat waste of rubble in the middle of the night is a precarious oddity.

April 11

IT IS A BRIGHT, cool morning. In full sunlight, the gravel plain rolling to the horizon seems cheerier. Two lone treetops emerge from a fold of ground about a mile away; they are literally the only features in the entire landscape.

Three Journeys

For the past few minutes, I've been hearing a rhythmic groan somewhere in the distance. It sounds like the agonized snores of a giant having a nightmare. Far to the left a speck has gradually taken form, inching across the horizon. It is amazing how even the smallest detail becomes an event in the desert.

Now the speck has increased to a finely traced miniature: a nomad on his camel moving across the plain. The snores and groans are the camel's usual ill-natured protest at existence.

After the icy mist and wind of the High Plateau region before reaching the Sahara, camping has been simple and pleasant. A few hours before sunset, I look for a patch of smooth flat ground. Dried mud or sand is best, but fine crumbly gravel will also do. Even the rough surface of the *reg* has occasional low spots where a bed of sand has been deposited by windstorms, and these make perfect campsites, the pure, rippled sand resembling a yellow pond in a plain of raw black stones.

Whenever possible I pick a spot out of sight of the track, so that my campsite will resemble a desert island: the tent, a folding table and chair, a can of water, my stove set up in the tent mouth away from the wind, the car caked with dust on the side opposite the sunset. Tea, crackers, a fluted glass of calvados, unfortunately lukewarm. Around the tent, strings of footsteps like garlands trailing over the sand. The garlands become more complicated as the evening approaches and I move from one task to the next.

Sometimes I take a walk at night, placing a light on the table to guide me back. At a distance of a few hundred yards, my island seems vulnerable and strange. Yet the feeling I have is not fear, but sumptuousness, luxury.

I don't have to worry about the view. The desert is all view. I feel like a fish moving its tiny bowl from place to place in the middle of the sea.

The freeze-dried food I brought with me is tasteless, but I find that doesn't matter very much. In the day's routine, I tend to eat quickly and simply. In addition to freeze-dried packages, I have plenty of canned sardines and tuna fish, some canned vegetables, powdered milk, instant coffee, lots of tea. Once I bought a piece of lamb in an oasis for an incredible price, and found it so tough I ended up sucking it for juice and tearing off shreds. When I spit the gristly chunks onto the sand, a black beetle surfaced beside them like a submarine, and dragged them away one at a time. On the other hand, excellent vegetables are plentiful in the oasis markets, especially onions, carrots, tomatoes, sometimes potatoes. Hole-in-the-wall bakeries produce French-style baguettes which are eery to see in the sand-paved alleys of a *ksar*, like surrealist flutes. The trouble with this good food is that it takes only a few hours for it to wither in the dry desert heat. The meal after passing through an oasis has a bit of substance; the leftovers are for the beetles, and on to sardines and rice.

April 12

THE WIND has gotten steadily worse since I left Aoulef this morning. About halfway to In Salah, the clouds of brown dust thickened until I was afraid I might lose the track, so I decided to stop the car and wrap a tarpaulin around the motor. I hope the generator doesn't clog up again. I'd hate to have to take it apart and clean it, something I've never done.

This gives me a chance to catch up on my journal. Two days ago, I spent a morning at the oasis of Tamentit, a few miles south of Adrar on the track to Reggane. Tamentit was an important Jewish settlement until it was destroyed in the late fifteenth century, and I was curious to see what traces were left of its former history. The usual swarm of boys descended on me when I got out of the car, and as usual, I picked the one who spoke French best and loudest to take me around, while fending off the rest. I asked him if there were any old ruined structures in the oasis. At first he didn't seem to understand me. Then I realised that, although he understood, he couldn't see why anyone would be interested in them. Mainly he wanted to take me to a shabby room in a mud hut where some mediocre artisan's ware was for sale; also he wanted to show me something he called the swimming pool. I managed to miss the swimming pool, but I bought a necklace of worked leather packets filled with charms against bad luck, called a *grigri*.

The palm grove is sprinkled with the vestiges of large buildings: a *ksar;* some ruined mud arches which the boy called the mosque. The present ksarians have a superstition about it. When someone passes by he drops a stone onto a heap, or places a flower or a piece of vegetable as an offering. The boy did it and advised me to do the same. I have no idea if the ruins I saw have anything to do with the destroyed Jewish town of Tamentit. They certainly cover a lot of ground. But would dried mud constructions last for 450 years, even as ruins?

In any case, there is no trace of former splendor in Tamentit. The *ksar* is a warren of little canyons and tunnels, with people crouched, or stretched out sleeping along the walls. There was an air of ill-health, even misery, in many faces. The draped bodies lying in the street with their knees pulled up looked as if they had died and shrivelled where they lay. The boy took me to see a jeweler at work in a minute cave of a room. His silver came from old French coins which he melted in a charcoal fire and beat into strips. It was arduous work, and the results were not very appealing.

The loveliest structure I saw in Tamentit was a magnificent marabout built on a mound in a wide, stony gulley. The understructure seemed to be the vestige of a former building, like the omnipresent ruins dotting the palm grove.

The mere presence of Jews in the Sahara is mysterious. No one seems to know when they came, or how. There are still Jewish villages in the Souss area of Morocco which trace their lineage to the Babylonian diaspora. But mythic genealogies are the rule in the Sahara. Nomad tribes often claim descent from the ancient kings of Yemen—an impossibility—or from various Old Testament figures like Jonah or Ishmael. So there is no way of knowing how much history is contained in the Jewish tradition. Another likely source of Jewish migration was the Cyreniaque in northern Libya. Under Roman rule, the Cyreniaque was a prosperous area with a large Jewish population. According to E. F. Gauthier, a nihilistic uprising of Jews took place in the fourth century A.D. The revolt was so savage that the Jews were said to have made body ornaments out of the intestines of the Roman dead. After the uprising, they fled, and where could they have gone to escape the Roman legions but into the "land of beasts," the adjacent Sahara desert? Arab historians remark that the most ferocious resistance to Islamic expansion into the Sahara in the eighth century came from tribes of Judaized Zenata Berbers led by a queen, the Kahenna. Wherever they came from, it is clearly Jews, not Arabs, who represent the first historic migration into the western desert from the Middle East. Their original numbers must have been small, for there are virtually no ethnic Jews in the Sahara, only Berbers converted to Judaism anywhere from 1,000 to 2,000 years ago. The last sizable Jewish settlement was in Ghardia in the Mzab. The Jews of Ghardia panicked in 1961 when Algerian independence was declared, and fled, mostly to Israel or to Europe.

Until 1492, the Jews of Tamentit are said to have been at the hub of Saharan commercial life. They were famous goldsmiths and merchants, and their location in the Touat was extremely important. The Touat is the jumping-off point toward black Africa. South of it is the Tanezrouft, a waterless desert almost 1,000 miles wide. North of it, a string of oases forms stepping stones to the countries of the Mediterranean coast. Tamentit was the key settlement, not just another dusty town in the desert, but a commercial center, wealthy, architecturally ambitious. It was destroyed during one of the periodic revivals of Islamic puritanism which enflame the desert now and then, this time as a reaction to the final stages of the Arab defeat in Spain. The Jews of Tamentit are said to have fled to Ghardia, where the Mozabites, a heretical Islamic sect, protected them while using and despising them, in a complicated relationship which lasted almost five centuries, only to be destroyed in several days by fear.

There are other mysteries too. Is it true that Jews continue to live secretly in the Sahara? I had read that a part of Timimoun was still inhabited by Jews who had been forced to convert to Islam in the fifteenth century but continued privately to observe Jewish customs. I walked up and down every alley in Timimoun without seeing a trace of any cultural anomaly. If there are Jews in Timimoun no one knows it, and the Jews are as black as all the other ksarians I saw.

Another mystery concerns the origins of a blacksmith tribe, the Ineden, who live with the Touareg in the Hoggar region of the Sahara. They too are black, and speak a language of their own, which is rumored to contain a number of words in the Hebrew dialect once spoken in the Touat. They are despised and feared by the Touareg, somewhat like gypsies in Europe today, but they supply all the Touareg's artisanal needs, and are doctors and goldsmiths too. Are the supposed Hebrew words in their language evidence that the Ineden are descended, however distantly, from the diaspora of Tamentit? Again, no one knows.

In any case, not a person I spoke to in the Touat had heard of its Jewish past. The contrast between the Cyclopian leftovers scattered in the gardens of Tamentit, and the wispy, hungry lives it now harbors, will haunt me for a long time.

Last night at Aoulef I went to a festival celebrating the name day of the prophet. The festival was held in a flat sandy square in front of a maraboutic shrine glowing faintly white in the candle flickers. Several dances were going on when I arrived, although at first all I could make out was a cluster of shapes and noises flitting about in the dark. One dance was held near the entrance to the shrine. A small circle of men and women clapped their hands in a complicated rhythm while they swayed gently around a drummer squatting in the center.

Meanwhile, in the middle of the square, a set of drums was

pounding a darkly joyous beat. Dozens of men in white turbans and flowing white jellabas started to shuffle in a circle, holding rifles upright in front of their faces. The dance is called the *baroud*, and is performed at virtually all festivals in the Sahara. It had been going on for some time when I arrived, and would continue all night, I was told, mounting to a climax time after time, as the men danced with increasing frenzy and delight.

More than a hundred men were chanting and shouting. Their faces glistened with sweat and their bodies moved up and down with short powerful steps while they gestured rhythmically with the rifles. The only light came from a storm lamp which a lone dancer carried on his head in the middle of the circle, so that the dance seemed to give off a smoky radiance. The men shuffled around several drummers performing a separate dance in their midst. Near the drummers, responding to them and also to the chanting, shuffling circle, were three wild-looking dancers making violent contortions with their rifles. The controlled savagery of the dance was exalting. The drums and dancers increased their rhythm in a stately crescendo until the climax, when all the men leapt toward the center, firing their rifles in a booming flash, which made everyone shriek with delight. After a short delay to recharge their rifles, they began again, and again.

I've been waiting for three hours, and the wind seems to be getting stronger. The Tidikelt Plateau is high and stony, so the sand isn't quite blinding. But the track is not well marked, and it would be foolhardy to drive in all this blur. The dust patches undulating toward me resemble yellow ocean waves. Yet, despite the delay and the gritty weather, I feel perfectly content, that's the strange part. A little tired from squinting, my skin cracked and dry, needing to urinate and not wanting to face sand in my eyes and in my fly. But happy. It's been that way during most of the trip. Even when I feel depressed now and then, it doesn't last, because, in a sense, there's no one to be depressed for; the dramatic element is missing.

Three Journeys

April 14

IT'S BEEN twenty-four hours and the sandstorm hasn't let up even for a minute. A few hours after stopping yesterday afternoon for coffee and the notebook, I lurched stupidly into a trap of clayey sand, while going down a steep incline. The car was stuck so badly that I had to dig out the entire bottom, and then jack it up to get sand ladders under the front wheels. It took hours of hard work, most of it lying half under the car in a swirl of dust and sand, while the sun drilled down at me. By the time I got the car out, my lips were caked with dried sweat and dust; I was trembling from exhaustion, and I was scared. I drove on for a while, sticking to the hard portions of the track, despite cavernous ruts, while I tried to clear my mind. I felt chilled and dizzy, and wondered if I might be suffering from sunstroke.

I decided I had better make camp right away, even though it was early in the afternoon. I stopped on a high gravel plateau, where the wind seemed less dusty. The plateau was rimmed by low chalky cliffs and mounds of pale grey rock. Had I been less exhausted, I would have found it lovely. Making camp was another ordeal. The wind blew the tent away before I could get it staked and weighted down, and I had to run after it. Even after I managed to set it up, the steady pounding of the wind made the sides of the tent flap deafeningly. I wondered if the nylon cloth would tear under the pressure. By then I was almost too tired to care. I collapsed on my unrolled sleeping bag and lay there for a long time. The wind usually dies down at sunset, but this time it blew all night, the flapping and gusting keeping me half awake. At dawn I stumbled everything into the car helter-skelter.

After only a few miles I drove off the track looking for some hard, level ground, and before I knew it I was stalled in sand again. The nearest hard patch I could see was at least a few hundred yards away up a slight incline. Every lurch with the sand

Against Emptiness

ladder got me the length of the ladder, about a yard and a half, and took me at least five minutes of digging. The wind was dusting and sanding. It is curious how calm one becomes at such moments. There was no one to complain to about discouragement and fatigue. It was going to take until nightfall to get out of this, maybe longer, and I was still shaky from yesterday. I stopped to organize my thoughts. This time I would dig for short stretches, and rest, and drink enough water, and I wouldn't take my shirt off. It might take a while, but eventually I would get clear. Luckily after half an hour a Land Rover passed by on the track, the only vehicle I saw all day. It towed me out in minutes, and I thanked the driver from the bottom of my heart.

By now I had visions of a shower, a room, some relaxation. But I wasn't finished yet. There were still thirty miles to In Salah, where I would join the main trans-Saharan track. The wind was rising again. The track came down gradually into a flat, sandy plain and soon the air was so thick with grit I couldn't see more than a few feet. The track had been covered by a layer of dust, which made it invisible. There were no markers at all. I should have stopped and waited. But In Salah was only a few miles away. I could already feel the water trickling over my body and the grit washing out of my hair. So I kept on, following the track mostly by feel: as long as I felt the regular jolting of the washboard surface, I knew I wasn't lost. After about an hour of this, I saw the most consoling sight in the world: a dark rim of palm trees bending in the wind, barely visible through the fog of dust. When I entered the grove, the yellow blur narrowed to a channel between rows of palms. It was an unspeakable luxury to be surrounded by actual shapes again.

When I think I almost made this trip without a tent! After a day of squinting and bright spaces, the tent is a consolation; I find that I need it, as one needs to be loved and hugged. Yesterday, when I was so exhausted, it would have made more sense simply to unroll my sleeping bag, crawl into it, and sleep, and leave it at that. But I wanted that green and yellow space stretched around me. I wanted to light my gas lamp when night fell, and read for an hour, and write a few words in my journal. Without the tent, I don't think I could bear the desert nights.

April 16

I SPENT A DAY wandering through the baked alleys of In Salah eating oily meals, and talking endlessly: a flood of choking congeniality. The main café and restaurant in In Salah is owned by an energetic black man who also owns a Turkish bath down the street and does odd jobs for the town government. He is so quick and busy, he reminds me of a nervous octopus. The café was filled with Algerian soldiers making believe the green tea they drank was scotch, and actually getting drunk. A couple of truck drivers sat in a corner with bloodshot wide-open eyes. In another were two English students with half the desert sweat-caked to their skin, and backpacks on the floor beside them. They were hoping to get a ride the rest of the way to Niger, but they didn't seem to be doing much about it. They sat in the café with an air of depressed patience, and waited. Every now and then, one of them muttered as if chewing thoughtfully to himself, and sat straight up in his chair with exasperating politeness.

The meal was a parody of French cooking: a few stalks of salad, a lump of meat the size of a half dollar soaked in fried oil, some stringy string beans, and an orange. The bottle of soda I drank cost half the price of the meal.

Aside from the palm grove and the government buildings, there isn't much to In Salah. The usual mud houses, and a marketplace with a struggle of merchants selling vegetables, a few battered jewels, and a sumptuous python skin from Niger. I tried to buy the python skin but got into an argument with the owner, who didn't seem to be bargaining fairly. I stalked away, and then felt pretty foolish, because I really wanted that skin.

In Salah is bounded on one side by dunes which have already

Against Emptiness

half buried a mosque. To keep the dunes from coming closer, wattle fences have been planted along their crests, like a strange hairdo. The fences cut off the plume of sand one usually sees streaming from the dunes on windy days. The blown sand gradually accumulates on the side away from the wind, causing the bulk of the dune to shift slightly. Since wind patterns in the desert are consistent over the centuries, isolated dunes move slowly in one direction, unless their movement is stopped by means of a windbreak.

I had been warned by travellers coming up from Niger that the first hundred miles after In Salah would be tough. Sand had drifted over the track, and the desert flats on both sides were dangerously soft for anything but a Land Rover. By now I'd begun to have enough of sand ladders and digging all afternoon to move a few miles. So I offered to take one of the English boys with me to help dig out and push. The other boy would follow on the weekly bus to Tamanrasset. There was no way I could take both of them. I would be carrying fifteen gallons of water, and twenty gallons of gasoline, in addition to all my equipment, and I didn't want to risk smashing the car's suspension on the rough Hoggar track, with almost five hundred miles between In Salah and the nearest stop to the south.

The last night I slept in the Turkish bath for a few pennies, picked up my hitchhiker before dawn, and started on the southward journey. It's a relief to be gone. In Salah was too crowded. I'd been away from people for so long, I found it exhausting to deal with them all day. My solitariness had billowed about me like a comfortable garment for two weeks. Suddenly among these English, Swiss, and Frenchmen, I was shrunk into a smaller, more talkative identity. Maybe it's just as well to have been called back to order for a while.

So far the track has been good, bad, and grotesque by turns. The English boy, Greg, helped me nicely out of the sand traps with some judicious pushing. In that sense, it's lucky I took him along. After about a hundred miles, I managed to get lost in a sudden dust storm, and swooped happily along a lovely blue-black valley for another hundred miles, following some car tracks. To my dismay, they led straight to a geologist's camp, where I was gently informed of the mistake, and had to go back, following my own tracks. At first I was crestfallen. But the valley was so beautiful: a dark undulating surface, with waterlike patches of sand here and there, bounded by cliffs with dark red veins running through them.

The idea of being "lost" began to seem comic as I swerved around patches of flinty stone and across hard sand flats, smooth and silent as ice. Why hadn't I accepted the geologist's invitation to have tea, and then driven into one of the tributary valleys and camped for the night in an unused stretch of rock? Instead, I had cursed, leapt into the car, and raced back along the valley, as if I had someplace to go when, in fact, I didn't, because I was already there: surrounded by the magnificent cliffs of the Adrar n'Anet.

I think my reaction was influenced by Greg's presence. He is doing his best to be helpful and quiet. But even his politeness is beginning to get on my nerves. Now and then I manage to recover my desert calm, sitting down to write, or, this evening, going off to take pictures while Greg, exasperatingly well-bred, set up the camp. I especially resent having to talk. When I don't talk, I feel as if I were immersed in a mysterious inner fluid which connects me to the mineral quiet, and the colors, and the motions of the air.

We drove through the Gorges of Arak today, and have now pitched camp on the edge of a gray, rocky *oued*. The washboard surface of the track had been smoothed for quite a distance this morning, and the driving was easy. Since then I have been forced to slow down to less than ten miles an hour by the most terrible washboard I ever drove on.

The landscape is becoming mountainous, but mountains unlike any I ever saw: black carved slopes, like enormous reliefs chiselled and thrown down over a vast space. The harmony of colors and shapes is unearthly. We passed a herd of camels being driven north

Against Emptiness

by two Touareg camelherds. They had been nineteen days en route from Tamanrasset, and expected to reach In Salah in five days more. Both men seemed noble and simple; not at all like the sly, noisy population of the oases.

Later, when we stopped for coffee, a robed man climbed up to greet us out of the *oued*. He spoke enough French, mingled with sign language, to indicate that he had been grazing camels for a month in a nearby valley, waiting for a truck which would take them north to In Salah to be sold. The man was marvelously elegant. I invited him to share our coffee, and we squatted in a circle, drinking with the loud slurping noise which people here make when sipping a hot drink. When I asked him if I could take his picture, he agreed, but asked me to wait a moment. He carefully rewound his turban, which was at least thirty feet long. He arranged his robe, shirts, and jewelry, and looked into a mirror which he carried around his neck in a leather pouch. Then he nodded to me gravely, and I took my pictures. As we drove off, he waved his arms in a broad, friendly goodbye.

The desert nomads exude a feeling of dignity and reserve which has not changed despite the misery of nomad life during recent decades. I find it very moving. The *oasiens*—their former slaves—are the ones who are benefiting from the Algerian government's economic effort in the Sahara: new schools, roads, electricity. But for the nomad it is different. His sources of prosperity—the organization of caravans, pillage, the sale of protection to travellers—have disappeared. More important, his moral advantage—the complex skill enabling him to master the distances of the desert—is now gratuitous and out of date since cars and trucks have enabled any amateur to possess mobility in the desert, a mobility superior to the nomad's but morally and spiritually neutral. The Reguibat of Mauretania are the only *grands nomades* left in the Sahara. Aside from the *azalai*, the twice-yearly salt caravan from Taoudeni, and smaller salt caravans from the Amadror region of the Hoggar Mountains and, farther east, from Bilma, the time of the caravan is over, at least for now. As a result, the nomad, who formerly was master, has fallen lower than his slaves.

Three Journeys

I think of the few Touareg I have encountered so far. With their veils wound arrogantly about their faces, they give a sense of pride and physical strength as they glide and roll across the open spaces like samurai warriors. Until now, the Touareg of the Hoggar have refused to work on road gangs or in the oil fields of the northern Sahara. Instead they spend their time visiting from tent to tent and exchanging elaborate courtesies. The Touareg make me think of the feudal classes of eighteenth-century France. Having lost their power to rule, they devote their traditional skills to a display of glorious uselessness, a kind of elegy which they perform as an answer and a dreamlike solution to misery. Perhaps it is a form of death wish; or rather, a desire to choose one's death, and not submit to it slowly, humiliatingly. The Touareg chooses to be no one, rather than become less than himself. The result is small, miserable encampments, sickness, a deep, deliberately incurable poverty.

April 17

TODAY WE PASSED from a landscape of low cliffs to the first steep thrusts of mountain, which tomorrow will become the peaks of the Atakor in the central Hoggar region. The contrast is startling between the rubble heaps of black rock broken down by heat and wind and the sculpted granite heights swimming above them. When a wiry bush or a wind-broken tree juts up now and then, it seems out of place, a flaw in the self-possession of so much stone and sand.

We rattle and shudder forward slowly, hoping to arrive at Tamanrasset intact. Every day has some incident. Today the extra gas tank sprang a leak. We had to stop and mop gas out of the car every few miles. Finally we siphoned off the remaining gas into an empty water can. I hope we can get it welded in Tamanrasset. If not, I'll have to use twenty-liter cans for the extra gas and carry less water. Tomorrow, Tamanrasset, hopefully in one piece.

April 18

THE ATAKOR was on my left about twenty miles away as I approached Tamanrasset. From a distance, the mountains resembled solidified explosions: a frieze of abrupt black peaks, veined with rose. They are often called lunar, because of their pure vertical height, unsoftened by the hug of vegetation. And it is true, their finely etched forms rising over a foreground of scree and sand seem to be anchored on another plane of existence. I hope to drive into them tomorrow, to visit the Assekrem Plateau, if the car can make it.

Tamanrasset is an astoundingly banal little town. It is hard to believe that 500 miles of spectacular desolation lead here. It has a main street, tourist shops, restaurants, an airport, and apparently very little life of its own. Nonetheless, it is good to have gotten here in one piece.

April 19

NOT SO MUCH in one piece after all. In addition to the leaky gas tank, which should be repaired by now, the plug of the oil reservoir has sprung a leak. A garage attendant tried to fix it last night, but he only made it worse. Oil is now oozing from the plug, and I've put a can under the car to catch it. Oil is precious around here. There isn't any for sale this week. The two cans I have were given to me by a helpful truck driver. I may have to wait for a new oil reservoir to be flown in, but I won't know until I take the car to the town repair shop and plead with the mechanic to look at it.

Well, the car problems have been solved for now. I have an oil plug improvised out of an oversized spark plug, and a resoldered gas tank. I just hope they hold. Obviously I haven't lost my talent for anxiety.

Something else has begun to bother me too. I've begun to feel frustrated by the limitations of my trip. Travelling by myself in a 2CV forces me to keep to relatively well-travelled tracks where at least some traffic can be expected to pass. For example, if I can't find another car willing to go as slowly as I must, the desert police won't let me go on alone to Djanet, and they're probably right.

Travelling in a single car, alone, it is simply out of the question to aim for the more remote portions of the Sahara. For that I would need two or three Land Rovers, and would have to arrange for gasoline to be sent on ahead by truck or plane. I would also need a first-rate mechanic, a crate full of spare parts, and a navigator. Yet being on my own has been such an important part of the experience of these past weeks. Greg was a sweet guy, but I'm delighted to be rid of him. There's no doubt, I'm caught in the paradox. To be alone in the desert, I've got to maintain a short, careful lifeline. To go any further would take an expedition: more geography, yes, but in a way, less experience.

April 21

I'VE PUT OFF WRITING for two days now, because I couldn't imagine words accurate or powerful enough to convey my experience of the Atakor and the Assekrem Plateau. I'm sitting now on the east edge of the plateau, next to the Père de Foucauld's stone chapel, overlooking the highest peaks of the Hoggar. A cold wind is flowing over the plateau out of the already approaching night. The peaks stand apart from each other in attitudes of solitude, each one thrust vertically, yet somehow peacefully, out of the hummock of fragmented waste at its feet. A bluish haze announces the coming evening.

The volcanic rage which flung these peaks to such a height has long since been exhausted. Time has thinned the rage to a barren climax, and then transformed it into an element of prayer. The peaks themselves, their vertical walls fluted and shadowed, resemble great cowled figures, inclined to the east.

The more I try to describe this extraordinary place, the more my words seem wrong, for the silence of the Atakor is the exact and haunting opposite of words. Looking out over the highest peaks while the sun lowers, and blankets of shadow wrap themselves around the stone feet of the mountains, I had the sense of peering inside a soul.

After the aimless mood of Tamanrasset, the desert was like a blade honed to an invisible cutting edge. The transition upon reaching the Atakor was sudden. After twenty miles of sand and scree, I was plunged into a spacious, vertical world.

I have never seen mountains formed this way before: each standing apart, its sides all but vertical with rough, organlike flutings, as if the stone had been marked by the resistance of the earth as it forced its way upward. The mountains seem unearthly because they are so subtly, almost humanly, formed. Yet, as one advances deeper into the Atakor, one realizes that the most intense unifying quality of these mountains is the absence of all visible life. Wherever one looks there is stone: wide mounds of scree with the peaks rising out of them; plateaus sown with glistening black rocks the size of heads and fists; *oueds* gray with sand and water-smoothed boulders. Every conception, shape, and color, every life-mimicking combination of stone. I thought of the legend of Deucalion and Pyrrha, who threw stones over their shoulders so that men would spring up in their place. They sowed busily in the Atakor, but the stones never turned into men. Yet the landscape, or stonescape, isn't barren. Deucalion's stones are strewn everywhere. At the sight of their immense passiveness,

Three Journeys

one almost believes that their time is still to come; that a life hums in the taut silence of the Atakor.

Two brothers of the Petits Frères de Jésus live permanently on the Assekrem Plateau in stone hermitages. They keep the chapel and a small museum devoted to the Père de Foucauld. I spent part of an afternoon with the brother Pierre, a young man who has been living here for two years. The other brother, Jean-Marie, whom I met briefly today, has lived on the plateau for twenty years. Pierre did not impress me as a man lost to the world. On the contrary, he was affable and well-informed about world affairs by means of a cranky transistor radio he managed to coax into life now and then. Obviously he likes people. He said that he thought a misanthrope would go mad in the desert. You've got to enjoy men, he said, to be able to abide such enormous solitude.

I was moved by the directness and simplicity of Pierre's faith. As I talked with him, I had the sense of a complicated man, an intellectual perhaps, who deliberately chose, and little by little forged, the simplicity of his present life.

Charles de Foucauld wrote, in the late 1890s, that his conversion to Christianity had resulted from his experience of the Saharan peoples living simply, without ostentation, in the presence of God. Their piety was so complete, it became the ordinary style of their lives. Because they lived in God, they did not require pomp and elaborate behavior; their faith was not a matter for solemn occasions. It is moving to observe a Moslem step discreetly into another room, or kneel without fanfare in a corner of the street, to say his prayer.

Pierre gives the same impression. He built his own hermitage in fifty days, he said, using the flat stones the plateau is literally paved with. It is about one kilometer from the chapel, overlooking a spectacular array of peaks. His hermitage is one of the most stirring places I have ever seen: a squat, square shelter, half disappeared among the rocks of the plateau; inside it a stone altar, windows decorated with line drawings from the Gospel, a bed, some books, a barrel of water. And prayer, a permanent prayer, interrupted only by small duties: greeting visitors on alternate days, making meteorological notations for the Algerian government, which pays the brothers fifty dollars a month to live on. Interrupted also by solitary walks in the mountain valleys, often for days at a time. But not interrupted at all. That is what impressed me most about the man: everything he did, it seemed, was done in the mode of prayer. It must be an extraordinary challenge to live in such a place. I don't mean the isolation, which is prodigious, but the need one must feel every day to exist on a level of wholeness and piety which the enormous mountainscape embodies so absolutely.

April 22

PIERRE TOLD ME about an excursion I could make on foot through a network of gorges and *oueds* near Assekrem. I started before dawn the next day and got back in time to see the sun set from the plateau. I must have walked twenty miles in a circuit which took me to the foot of a great domed mountain, named Houl in Tamachek, meaning "the heart." To reach Houl, I scrambled down a narrowing canyon, its floor littered with boulders. A ribbon of bushes and even some flowers were scattered along the dried bed of the *oued*. All along the way I found camel and goat droppings, some quite recent. Wherever the canyon widened, there were traces of old Touareg encampments. Small spaces had been cleared of rocks, and were rimmed by low dry-stone walls. In some places, there were as many as a dozen of these tent-sized spaces. In fact, the further I penetrated into this extraordinary stone wilderness, the more signs of human life I discovered, all of them concentrated along the narrow lifeline of the *oued*.

About three miles down the canyon, a small *guelta* filled with water took up the width between the canyon walls. After miles of millennial drought, it was uncanny to turn a corner and find, cradled in a white stone basin, a pool of clear water.

Pierre had mentioned a waterhole in another *oued*, some miles further on. After much searching I found it: a hole about twelve or fifteen feet deep in the floor of the canyon, its sides shored with branches and rocks. A series of ledges formed steps down to a shadowed crevasse at the very bottom of the hole. In the crevasse, under a crust of yellowish-green scum, was the water: crystalline, pure, and ice cold. According to Pierre, eight millimeters of rain have fallen in nineteen months. Yet here, under a rock ledge, the trickles of rain had filtered and remained. I gathered some twigs and made tea, Arab style.

Early in the afternoon I entered a gorge worn deeply into the volcanic substance of the mountain. Pierre had told me I would find Tifinagh inscriptions carved on the walls of the gorge, along with crude engravings of animals. The quiet of the gorge was intensified by the noon sunlight. The sound seemed to be drained even out of my footsteps. After about a mile I began to see columns of symbols scratched with great care into the walls; they resembled hieroglyphics or mathematical signs and were distributed haphazardly throughout the canyon. One house-sized boulder had been covered with them, crowded against each other, some even superimposed, in many different handwritings. In places the inscriptions were interspersed with animal engravings, mostly of antelopes and horses, an ostrich I think, and one childlike representation of a camel. The most beautiful engraving was of a galloping horse. It was stylized and symmetrical, unlike the crude attempts at naturalism which characterized many of the engravings. At least one deeply incised figure placed high on the rockface of the gorge represented a man mounted on a horse. Because the animals in the engravings are for the most part still found in the desert, and because it is known that horses were used by the Touareg probably until a century or so ago, I'm virtually certain that the engravings as well as the inscriptions were done by the Touareg, and are not prehistoric, although they could easily be as much as two thousand years old. The present day Touareg apparently have special regard for them because there are no campsites in the gorge. It was the only part of the mountains I walked through where this was so.

Against Emptiness

The Touareg inscriptions look a little like graffiti: bits of language scattered playfully, and maybe a little arrogantly, over the empty surface of the desert. It is an extraordinary thought: city children spraying their names on subway cars and Touareg women scratching theirs on a canyon wall in the Sahara play the same slightly mad game with empty places. Could it be that language and civilization themselves are forms of graffiti scrawled playfully on vacant slabs?

A few centuries ago Europe apparently forgot an important cultural-geographical fact: that the Sahara Desert, not the Mediterranean Sea, formed the true southern boundary of our civilization. The *limès* of the Roman Empire was traced on the far side of the Atlas Mountains, where the African and European continental plates collide, creating a sudden limit to the desert. A hundred miles to the north of it, Saint Augustine smelled the Saharan wind, and watched its brown glow dulling the heavens. Further to the east, Moses forged the stubborn will of his people in the desert, and Christ met the devil in the desert. The ascetic passion of Christianity was enacted among flinty hills and deep wells, date palms and camels: the Thebiade, and the deserts of Nitria. Saint Anthony and Macarius were crusty-skinned desert dwellers; Saint Jerome cleared his mind in the desert. Mohammed was cradled in the desert.

The great urban centers of antiquity—Babylon, Memphis—were phenomena of the desert, counteracting its influence with the monumental works of civilization. Their water systems, cisterns, and irrigation canals required precise measurements of time and promoted thoughts of time's vacant opposite, eternity. They gave rise to the massive structure of political absolutism and the playful structures of human will which were Babylon's gardens and Egypt's pyramids. In the realm of the invisible, the most imposing of all absolutisms arose at the desert's edge, as an answer to its inhuman emptiness: the vision of one God, monotheism.

The pyramids, the holy cities of Yucatan, the cave paintings of Lascaux are forms of the same joke played on emptiness. The Sahara was the abyss which taught us the joke, as were the bulging rock-faces of the caverns in the Vézère Valley and the scaly scrub lands of the Yucatan.

April 23

I WOKE UP in the middle of the night last night feeling sick to my stomach. A group of Frenchmen had arrived at the refuge in a Land Rover and invited me to have dinner with them—*boeuf en daube* in a can. After these weeks of austerity, the meat and rich sauce must have been too much for me. I spent the night stumbling around on the hillside, throwing up and freezing in the high-altitude chill. In the morning my head ached, and I felt dizzy. The track swooped and bent at incredible angles. The ridges of the washboard surface jolted the car mercilessly every time I gathered speed to climb fifteen-degree inclines which mounted up, around, and further up, before they hairpinned their way along a cliff edge and down into the next valley. The landscape became nightmarish and hot.

Late in the morning I noticed a cave mouth on a slope high over the track. I had a feeling it might contain inscriptions or etchings and decided not to let nausea and a headache stand in the way of curiosity. I stumbled up the slope, kicking showers of round stones down behind me. The cave had recently been used as a shelter by Touareg nomads; the walls were blackened by smoke, and the floor was thick with dried goat droppings. But along the back wall I found a gallery of partly effaced drawings, overlaid with soot: red-outlined shapes, like the wall drawings of the Tassili N'Ajjer of which I've seen photographs. Only one human figure was clear enough to identify with certainty. The drawings are probably prehistoric, since no Tifinagh inscriptions were to be found in the cave. I looked on the ground outside for chipped flints and other manmade remnants, and found a curious fragment of stone with a smooth beveled edge, probably a piece of a stone jar. By then my head was throbbing violently and I eased my way down to the car.

Against Emptiness

By midday I was so exhausted and hot, I stopped the car on the edge of the track. The sun was vertical, and there was no shade anywhere. I felt too sick to be hungry, and too tired to put up a sun shelter, so I rolled myself under the car and went to sleep in the dust. I woke up an hour later feeling more human, especially after crackers and tea.

During the early afternoon the landscape began to change. I had left the fluted peaks of the Atakor behind, and was now hissing through long washes of dust and gravel, between the familiar rubble heaps of scree. The *oued* the track ran through was speckled with yellow bushes and desert acacia trees, their main branches broken and trailing on the ground. Because of sparse rainfall during the past few years, Touareg goatherds have been breaking off branches to put them within reach of their flocks. The trees will probably not recover. Year by year, the Hoggar is losing its slender privilege of vegetation.

It is hard to believe this region was called the garden of the Sahara by Arab travellers during the Middle Ages. Man, not climate, has been the destroyer. He has thinned out and eliminated forms of life which were already leftovers from a moister period in Saharan biography, what naturalists call fossil flora and fauna: trees, grass, animals which have survived by means of strenuous adaptations, learning to thrive on hunger and scarcity, like the small crocodiles discovered a few decades ago in a permanent *guelta* of the Tassili N'Ajjer, hibernating in the mud. The crocodiles ought to have been extinct for 6,000 years. But the orchestrator of natural selection had reduced their size and their hunger and slowed their breathing, and they crawled in the muddy reaches of the *guelta* like shards of life from another age.

As the Sahara has continued to dry out during recent geological time, even man has become a form of fossil fauna, his presence maintained by a network of improbable circumstances: colonialism, the discovery of the desert's mineral wealth; before that, the sporadic energy of trans-Saharan trade, salt mines to satisfy the black African hunger for salt; a finely balanced interdependence of nomads and *oasiens*. All of these elements are as precarious as the ecology of the acacia tree in the Hoggar. A small change in the human weather: political chaos in Algeria, a war, a world economic crisis, and man too may disappear from the Sahara, like the lion and the Carthaginian elephant before him. Or he will hang on in enclaves of adaptive technique, like the ostrich, the antelope addax, or the Saharan cobra, the almost mythic *naja*, which burrows yards deep in the sand to find pockets of moist cool, and dances to the bizarre flutes of the snake charmers of Marrakesh.

After a while I began to pass flocks of black and white goats guarded by Touareg women. One time, when I stopped to take a picture, a woman waved to me and came over to where I was standing. Her face was deeply wrinkled, giving an impression of energy, but also of calm, resignation. The woman was black and dressed in black. Walking toward me over the light-bleached gravel, she seemed almost supernaturally intense. She came up and silently looked me over, neither friendly or hostile. She wanted water. I got my water can out and poured a cupful for her. We squatted on the sand facing each other while she drank, and then I poured a cup for myself. As I watched her black face with its thin, wiry energy, it occurred to me that no Arab woman I had come across in almost a month had anything like her presence.

I had read that Touareg women have a great deal of personal freedom. They are revered as transmitters of legend and teachers of the Touareg script, Tifinagh, which I had seen scrawled on the canyon walls of the Hoggar. Arab women, on the contrary, are virtually invisible. They are veiled, of course, but it is more than that. They hold themselves invisibly. They stand, walk, and sit as if their presence were an error which they correct by erasing themselves before one's eyes.

This Touareg woman wandering in the desert after her goats had a sumptuousness of identity which she held in check with grace. A few minutes later an adolescent girl ran up to us. She was the woman's daughter, it seemed, and she got down to business right away. Before I knew it, every loose item she or her mother possessed was offered to me for sale: a shell necklace, woven baskets, tin jewelry, bracelets, beads; everything but their clothes. I found it disconcerting, but I knew of the recent Touareg tragedy. The drought in Niger and Mali had destroyed the large camel herds grazing in the Sahel, many of which belonged to the Kel Rela tribe of Touareg in the Hoggar region. Traditionally the Kel Rela and their vassal tribes lived in the Hoggar with their goats, keeping their camel wealth hundreds of miles to the south where pasturage was more plentiful. But now they had nothing; their wealth was dead or dying. For the first time in recent memory, the Touareg were selling their personal jewelry, their *takubas* (swords), and their traditional home implements to travellers, to keep their nomad existence alive. The alternative was starvation, a mediocre government dole; or else they could settle near Tamanrasset and hope to find some sort of job, a slim hope indeed. I made up my mind to buy something without too much bargaining, and settled on a pair of wristlets which the woman took off and gave to me. They were delicately engraved with abstract patterns, in a light, silvery metal, probably aluminum.

The daughter was shy and flirtatious at the same time, but she didn't lose track of her idea, which was to sell me everything they owned, or if selling wouldn't do, then to barter. In no time she was dancing around my car pulling at every loose item: old towels, shirts, a plate, a ballpoint pen, a frying pan, a flashlight, some paper. The girl was possessed by a sort of crazy gaiety, while her mother sat on the sand, looking on impassively. Yet as soon as I had waded through the girl's antics and had gotten back inside the car, both women came up, smiling, and waved goodbye, with an expression almost of melancholy on their faces.

I reached Tamanrasset late this afternoon. After the splendor of the Atakor, it is strange to be seated at a café table, waving off clouds of flies, while the life of the streets ambles by: veiled Touareg men moving past with intense gliding steps; Land Rovers shrouded in dust; tourists fresh off the plane; local men walking or sitting around café tables. One never sees anyone doing anything alone. Companionship seems to be half of life for Saharans. Men often hold hands, and they touch each other constantly. The one thing about my trip that seems utterly strange to people here is that I'm doing it alone. At first this frightened me; I thought they were referring to the dangers of lone travel in the desert. But in fact they don't understand why anyone would want to do anything alone, even walk down the street, much less drive for days along empty tracks.

April 24

THIS WAS a sinister day. As soon as I left Tamanrasset, the sky became a white blur which elsewhere would mean rain, but here means only dust, closing down the horizon and stifling the sun without diminishing its heat. The landscape was oddly claustrophobic. The hills near the track loomed in translucid heaps, as if the opacity had been sucked out of them by the dust.

The dimness of the day became horribly precise shortly before dusk. A dead camel lay in the track, its body grotesquely twisted and bloated. Its stench drifted after me for a long time as I drove slowly over the chewed-up surface of the track.

I got bad news yesterday in Tamanrasset. I have to reach the Moroccan border in a hurry, or run some unspecified risk for being in Algeria without a visa. No chance to visit the Tassili N'Ajjer and the prehistoric wall drawings, or to continue on into Niger, or anything else, except to drive out, and hope my car doesn't break down, because they apparently don't make allowances for such things. I'm either out in time, or not, and the not is anything from jail, to a fine equalling all the money I have, to nothing at all. I can risk it if I like, but I won't. I don't like to admit it, but starting back seems almost like the right thing to do. I've begun to feel worn out, morally speaking. In particular I'm increasingly worried about the car. It's become an obsession. If only I were a better mechanic and had brought more spare parts with me.

Against Emptiness

Yesterday morning I went to the border police in Tamanrasset to have my visa extended, as I had been advised to do by the Algerian consuls in Washington and Paris. The border official was a thin, gasping man with eyes like broken glass. He took one look at my passport and accused me of being a spy, apparently because I had the bad luck to be an American, and also a writer. The logic was mad but if I wasn't careful, the man hinted, it might lead to prison. To complete the syllogism, all he had to do was call in a couple of soldiers from the hall. I tried to argue with him: "I'm a poet, not a reporter." "I'm not writing about Algeria, I'm writing about deserts in general." "I'm not really writing a book at all. This is a personal trip." "What could I spy on anyway out here in the Sahara?" His response took the form of an insane stichomythia. To everything I said, he answered: "*Oui, précisement*," with increasing venom which I unwisely ignored until a last outburst shut me up.

"For three years I helped the FLN fight for Algerian independence," I pleaded, "I risked going to prison, I risked my life; and this is how I'm treated."

The official glittered his broken-glass eyes at me, and hissed: "*Ici, on ne fait pas de la politique. C'est formellement interdit.*"

He leapt to the door, locked it, and turned to stare at me.

I sat very quietly, while he tried to decide whether or not he was angry enough to spend the day filling out the sextuple forms which accompany all official acts in Algeria, including putting foreigners in prison. To my relief, he decided in favor of laziness, and I got out of the office quickly, though without a visa extension. I have five days now to reach the Moroccan border, which is almost 1,800 miles away. My car has begun to burn oil at a frightening rate. On the way out of the Hoggar I began to notice a throaty noise in the motor, and less power I think, probably a loss of compression. A good part of the way will be over bad track at ten miles per hour. Another part over twisting mountain roads full of potholes and doubtful fording places. The white pall hanging over the track today almost surely means there will be sandstorms farther north.

Three Journeys

Despite my disappointment, I spent an extraordinary last evening in Tamanrasset. Toward the end of the day someone I'd met in a café accompanied me to a cluster of reed huts in the desert just beyond the town. The huts belonged to an Ineden blacksmith clan which specializes in making Touareg jewelry. The hut we visited contained a number of young helpers gathered around two sleepy-eyed, patriarchal men, the master jewelers. The jewelers bent cross-legged over anvils made of car axles and used tools refashioned from ordinary pliers and hammers. Their source of heat was an intense charcoal fire set into a hole in the ground and fanned by a goatskin bellows connected to the fire by an underground tube. The blacksmiths were making decorative sword handles, daggers, and a variety of Touareg jewels, each one designated by a traditional name. As in the Touat, the silver was melted down from old French coins.

The movements in the hut were quiet, almost limp. Jewels materialized out of a small chest for me to look at; tea was served; the metal work went on lackadaisically, as if its necessity resulted from the straw odor of the hut, from its particle-filled air and its drowsy comradeship, rather than from anyone's willed intention.

Like most Saharans the jewelers were black. I wondered about their supposed Jewish origin. But I had no idea what sort of questions to ask, or how to ask them for that matter. We had no more than half a dozen French words in common, and spent most of our time smiling and waving at each other.

I bought a magnificent silver pendant: flat and square with a geometrical design raised in bas-relief. I also bought a Touareg amulet called the *tineralt*, or *croix du sud*, a sumptuous variation of a cross the existence of which has caused people to wonder if the Touareg, in their far past, had not been affiliated with the Christian populations of Egypt and northern Libya.

When I got back to town, I was told there was going to be a *fête* on the outskirts of Tamanrasset to celebrate the vaccination of a young boy. The boy's father came from the Touat, and the *fête* was being given especially for his compatriots from the Touat who worked in Tamanrasset. I followed the drum beats from street to street until I found the place. Several dozen men sat packed together between the walls of a narrow alley, swaying ecstatically to the drumbeats and clapping their hands in a complicated rhythm. The men were huddled so closely together in the shadowy light, they resembled a many-armed creature undulating on a sea bottom. The host leapt up to greet me, and guided me along the wall to a place inside the alley where I could sit on the edge of a blanket. As I sidled past the clapping and swaying men, I passed an open door out of which poured a gust of warm air filled with rustling. I looked behind me into a corridor packed full of women, or rather of smiles and eye-gleams, which were all I could make out in the dark space. The women were not permitted to sit with the men. By crowding in the open hallway, they preserved the fiction of not being there. All evening trays of tea and crackers emerged from the hallway as if by levitation.

Outside, the men joked, shouted, and squeezed each other. During each chant, a leader leapt to his feet and made comic gestures which culminated in a burst of exaggerated clapping, and then, without warning, an abrupt silence. When he gave the signal, all the men stopped in mid-clap. If anyone added a clap too many, he was laughed at and pounded black and blue with bearish affection. I tried to figure out the clapping rhythms, and occasionally succeeded well enough to join in, paying attention not to be caught out at the end. The evening was filled with exuberance, and it gradually dissolved the anxiety of my morning encounter. Because I had to start early in the morning I left shortly after midnight, though the *fête* was going to continue until dawn.

April 25

THE DUST HAZE has thinned; an edgy breeze is flowing out of the sun, which is perched on the horizon, already gathering strength. My camp is on a plain of hard brownish sand, its surface rippled gracefully by the wind. The plain is rimmed by a series of domelike hills: several of scree resembling tumuli, they are so perfectly hemispheric; others, heaps of smooth boulders of all sizes, the interstices filled with sand, so that the hills look more like dunes inlaid with boulders. The mood of the desert changes so quickly. Today the sand- and rockscape has a crisp inviting air.

My slow-motion dash to the border continues.

April 28

GHARDIA is the first place I've been in a month with a telephone connection to the outside world. I just tried to call my wife from the Hotel Transatlantique, but there was no answer. The distance between us frightens me suddenly. I have no idea if she is still in New York, or has gone to Paris, or Venice, or if she has done something else entirely. For a month the Sahara formed a seamless environment. My previous life was like a puzzle put together out of remembered but inanimate pieces.

May 2

THE DRIVE out of the desert was like tunnelling through Armageddon. For 1,000 miles the wind wielded its blade of sand and dust. The surface of the desert moved with terrible leisure: sulfury yellow, mingled with gray, and an oddly tender rust, developing in serpentine undulations out of the livid distance of the horizon. Shortly after leaving In Salah, the lid of sand hissed shut. Awesome boulders swam through it at intervals, worn by millennia of wind into antediluvian shapes. In the rusty shadow of the storm, they resembled great shellfish stranded on the floor of a dried-out sea, its water replaced by a new element: dry and abrasive, in which you might drown as easily as in water, should you be foolish enough to open the car window.

Against Emptiness

By early evening, low dunes began to block the road. My windshield was too scarred and dusty to see through. I blundered over, swerved around, and luckily did not bog down in them. The next day from El Golea to Ghardia was only a little better. Here I could see the road; the light was a shade less grim, the sand less personally enraged at the passage of this noisy heap of tin and skin. Nonetheless, by the time I reached Ghardia, the relentless rubbing of what I do not dare call weather had worn my nerves to a smooth nub and destroyed the inner workings of the motor. I managed to convince a mechanic to spend the day taking the car apart and putting it back together again with a new set of pistons and cylinders. I had been reassured about the remaining few hundred miles after Ghardia: there would be no sand. If the wind blew, it would be the sort of wind I remembered from another life: invisible, composed of air, not moveable earth.

The next morning the wind did blow, a monstrous, freezing wind driving directly into the teeth of my car, and slowing me down to thirty miles per hour. And it was as full of dust as ever, occasionally so thick I had to stop until I could see the road again. Patches of sand poured across the road in opaque streams two or three feet high.

I stopped every few miles to clean the air filters, in a numb struggle to save my new cylinders. Washing an air filter in gasoline in the middle of a freezing sand storm presents interesting problems, among them how to keep the wind from tearing the hood of the car off when you lift it; how to see what you're doing in the gloom of driven dust; how to keep the filter from getting even more clogged than before, as you fumble to insert it into its fixture; how to keep the basin of gasoline from lifting up and spilling over you. All of these problems to be faced not one after another, but simultaneously. After a few hours of this I began to bellow at the top of my lungs in disbelief. It seemed to me that these sandstorms were defying all the laws of personal drama: they had no climaxes, no crescendos, no releases; only a permanent, unrelenting crisis, which rolled over me as if I weren't there. It was unartistic, and it was inhuman. I should add that the motor of the car has shown signs of giving out yet again. Car motors are a delicacy the Sahara consumes with great relish.

When the road climbed at last onto the High Plateaux, and bald patches of green which one would normally call impoverished had begun to appear, the wind became merely fierce in the old familiar way; Armageddon turned into what is normally known as bad weather. I had left the Sahara, or rather I had been swept out of it by the great swirling broom of the wind.

May 4

I'M BACK in the world again. Francine is sleeping a few feet away. A strip of ocean sparkles across the street from the hotel terrace where I'm having breakfast. I spent the day yesterday devouring French and American newspapers, old *Time* magazines. Tangier is an actual city, full of crowds and traffic. The circle has groped its way shut. I'm inside it. The desert is the space beyond it which I still cannot measure, although the space is not quite so abstract as before; on the contrary, it has become cluttered with images and memories.

This reminds me of a confusion which the experience of the desert hasn't cured. For a month now it hasn't seemed to matter, but I've begun to remember and be puzzled by it again. I still don't understand why this trip has been on my mind for so many years; and why, after the wind, the sandstorms, and the anxiety, I want to go back again.

Maybe the idea of the desert crystallized an idea of myself: a suspicion of personal emptiness which all my talking and my anxious attempts at charm surround and decorate, but don't penetrate or even come close to. Maybe I simply wanted to lose myself in a world which was the dialectical opposite of the one I knew. In my mental geography, the desert and the city are bound up with each other. The Sahara, mythically speaking, was created in Europe by men—Charles Doughty, T. E. Lawrence, the Père de

Against Emptiness

Foucauld—who dreamed of what they didn't have: space, the horizontal abyss.

It's no accident that Arab cities are the most urban of all cities. Their noisy markets and crowded, narrow streets undermine the very memory of space. Arab decorative art is crowded with details; it is compulsively, almost manically, elaborate. Even the tiny desert *ksour*, with their crenelated defenses and high-walled alleys, are not villages, but enclosed fragments of city.

Monotheism is not the natural religion of the desert, as Renan suggested. It results from an intense dialogue between the desert and its neighboring cities. In a sense, we are all neighbors to the desert.

There is something else, too. When Doughty went to Arabia in the 1800s, he claimed somewhat grandly that it was to revive the expressive possibilities of the English language. Well, in a sense, I also went to the desert to solve a problem with language, although not as Doughty meant it. Perhaps I can put it this way. It's possible to think of language as the most versatile, and maybe the original, form of deception, a sort of fortunate fall: I lie and am lied to, but the result of my lie is mental leaps, memory, knowledge. Portions of world are caught in my psychic net. I become human, and increasingly more human, because the acrobatic gift of my lie turns it into a truth of another sort. If "each man contains all the possibilities of human nature" (Montaigne), it is because language's acrobatic lie has thrown them in a busy dance about his ears.

But sometimes I get tired of lying and being lied to, fortunately or not. I want to be face to face with something I can touch. The aristocrats of old France declared: as for living, we let our servants do that for us. I'd come to feel that way too: as for living, I let my words do that for me.

It has been said that if a man lived his normal life for only a day, and then was locked in a room, the experience of one day would suffice for him to become a great philosopher and writer: the world exists in a grain of sand; each person, each lived hour, contains all the possibilities of human nature. But the locked in philosopher ought to remember that, after all is said, he has been out in the world for only a day; that the expanding links of his net may compensate for his lack of experience, as sharpness of hearing or touch compensate for blindness. Nonetheless, every day but one of his life has been spent in the same room, while the enormous whispering of the world has gone on beyond his reach. He ought to remember that his ability to experience personal truths is a counterpart of the poverty which has become a condition for his existence. His wisdom is real, but so is the poverty.

For almost twenty years I've spent my best hours thinking and talking about books. It seems, at times, that I have justified my existence because of the books and poems I myself would write. Having lived for a day in the world has given me plenty to think about. But I've been distracted lately. I've begun to hear an enormous whispering, as if the room had gotten smaller, as if something new and indispensable were taking place in the world beyond my reach. I want to find my way outside to spend a second day, maybe a third. I probably couldn't stand much more than that. Having gotten to know the limits of my room all too well, I've been shaped inwardly by its properties, so that if, at times, the room resembles a sort of prison, I've got to admit I like my prison, and will be content to come back to it when the time comes. I've come back to it already.

But meanwhile, for the strangest of days, I've navigated alone in the desert in a tinny car, busy with the small problems of survival, each day's preoccupation more powerful and more ordinary than a poem: how to dig the car out of sand when my muscles and ingenuity are the only nonmineral resources within a hundred miles. How to get a tent up in a sandstorm. How to sleep in a solitude which is denser and more solid than the earth's crust, a solitude stretching for miles in all directions, as exterior to my double handful of brains and thoughts as a meteorite is to the patch of ground it vaporizes.

I have travelled into the heart of a word, and discovered a

reality which exploded the word into its elements. For the desert I travelled in was not an ascetic ideal; it was not an inhuman border enhancing the features of the known world as a dark outline enhances the expressiveness of a color. The desert was a place, impure and complex, eluding definition. It provoked me to have new eyes and a new brain, while forcing me to recognize that my eyes, my brain could not be made anew; that the used familiar ones would have to do. But sometimes they didn't do. Even after the appearance of ordinary events had been stripped away, and I advanced inside a geological fantasy so gripping and ice-clear that it pulled me out through my eyes into its enormousness, all too often a struggle of anxious thoughts distracted me from my experience, and the desert blurred past my self-preoccupation.

I think I decided on the desert because it was the easiest place for me to go. My idea of the Sahara corresponded to an inner dryness; it resolved an emptiness within me, as one resolves the blank emulsion of a film into forms sharp with color and shading, foreground and background; as though a window had been opened in the fabric of the world. The Sahara, I guessed, would refract into diamondlike detail an intuition I have sometimes had: that the elements of my character, my desires and my anxieties, the intricate repetitiousness which is the room the blind man lives in, possessed a submerged counterpart, like the minute reversed image in a camera lens. My invisible emotions were inscribed in it as a scarab is inscribed in a drop of amber. Trapped in this spiritual amber, they possessed the fine, hairlike detail of a molecule under an electron microscope; they possessed everything but motion. I imagined this miniature to be the well of concentrated principle which stormed, suffered, and radiated into the world as me, but not really me. The "me" was sunk out of view, dense with concentrated light. I supposed that this was what was meant by the "soul."

The Sahara would be an immense analogue of this "soul." It would correspond to it, as in the old cosmology grass corresponded to the stars, rivers to the circulatory system of the body, and man to the earth itself. In the Sahara I would exchange dimensions. I would enter into the miniature through the medium of vastness. I would approach the concentrated brilliance at the heart of my life by entering into a terrain of intricate lifelessness. The "storm" of ordinary realities would become a scarcely remembered pinpoint enveloped by the dazzling changelessness of the "eye," which would reflect back to me an unrecognizable bundle of nerves and memories, a "me" moving wholly inside the diamond.

My trip, of course, did not turn out in quite this way. The desert was not the interior of a soul. I was not a mobile Robinson Crusoe crossing over a sea of rock. More often than I would have liked, shabby towns interrupted the emptiness, full of dust, made of square graceless buildings. Towns which displayed all the secondhand vestments of civilization: a rusty gas pump; bistros called the Café des Amis, or the Gargotte du Sud, resembling disaffected outhouses. And in each town, under a coat of fresh paint, behind a row of gleaming Land Rovers, the local offices of government, which gave the impression of having nothing to govern, their schedules a puzzle I never solved. Above all, in the oases, I encountered misery, chronic illness, undernourishment. Saharans have always existed at the threshold of disaster. But I had imagined something more spectacular: famine, fabulous sandstorms, bandits. Instead I found what Conrad once called a flabby little devil, except this devil was languid and thin, his eyes puffed into gleaming slits, with scabs of trachoma and conjunctivitis at the corners. He had a way of coughing and wheezing from the deepest hollow of his chest, a dry, matter-of-fact sort of cough, as if the coughing and wheezing were not symptoms of illness, but a condition of life, as fundamental to it as breathing was. The few nomads I met were leaner, but healthier. In the oases, apart from a small number of merchants and functionaries, everyone existed in a sort of physiological slow motion which measles, scarlet fever, flu, or some other epidemic that elsewhere would be merely inconvenient brought to a debilitating and often deadly halt.

Against Emptiness

In places I found paved roads, barely wide enough for a single car, which seemed to have been spun rather than built over the salt flats and the slag heaps of weathered rock. To be sure, the roads vanished at times under drifts of sand into which one sank helplessly, or else broke axles on when wind had compacted the sand into stonelike undulations. Nonetheless, the roads had an extraordinary effect: they miniaturized the desert, as if I were seeing it through a diminishing lens. Without its cutting edge of danger, which intensified the space and sharpened the remote masses of rock, the desert became boring; its uninterrupted size became flabby and unvaried.

Above all, there were people: those whose gestures and faces eluded me, because we had no language in common, like emanations of the desert which I could never wholly grasp. White-robed, glistening black faces, long shuffling steps; nomads who seemed to be exuded by the rocks wherever I stopped the car, coming up to me and standing silently, sometimes sharing my tea, but always standing or squatting; not waiting for anything, not impatient, as if this meeting in the desert were the precise form their solitude was meant to take on this day. These encounters unnerved me. I could never make my time correspond to theirs. Even when I understood, finally, that waiting and expectation had nothing to do with it, I waited and expected, gradually becoming impatient, and then angry with myself, for I suspected that I had been unequal to the timelessness which these men possessed in their lives and were now offering to me.

More often, there were people I could talk to: storekeepers, government officials, school teachers, truck drivers. These men lived in the oases. Their knowledge of French was a sort of insignia they were happy to display, for it signified their superiority over the desert. Among them, the government officials were the most tormented. Very often they came from the north. No one, I discovered, resents the desert and fears it more than a northern Algerian, to whom the Sahara usually represents a failure in his career and an exile, but more deeply, an atavistic threat he can never forgive. He takes his revenge by administering with a remote hand and smiling ironically at these "primitives," of whom his ancestors had lived in awe.

The ones who broke the spell most thoroughly for me were the other desert travellers I met: the more timid ones who hugged the paved roads, and exuded a spiritless ambiance whenever I encountered them; the braver, or merely richer ones, who cavorted around in their Land Rovers, winding themselves up for a mad dash to Tamanrasset and back again, or a madder dash all the way across into Niger, rarely slowing down long enough to see out of their own dust cloud. They made my own journey seem vulnerable to me. When I encountered them in the oases, or stopped and talked with them on the track to Tamanrasset, I was forced to recognize that I had not travelled through any looking glass; that the desert would not, of its own nature, contain me in a spiritual medium which was natural and objective; that even here I would have to supply eyes to see with and a brain to know with. It would not be enough simply to travel in the desert. I would have to give it an existence, I would have to model my interior spaces according to its prompting. When I was not equal to the task, the desert would be replaced by dull dusty spaces. It would become a desert in another, more terrible sense.

It is amazing that I was not prepared for this. But then, my idea of the trip was amazing on more than one count. From beginning to end, I experienced it in purely mythological terms. While in the desert, I struggled to remain inside the myth. I tried to harvest the myth from the desert itself, extracting it from the densities of rock and the sandstorms and the harassing wind; from the palm fronds and the intricate gardens which walled out raw space with a culture as intense as the arabesques of Moslem architecture.

The Sahara drove itself against me physically and demanded a physical response. But, as it turned out, I had not come there to be "in the world" at all. The enormous whispering I had listened to had been something else entirely. It had caused stabs of insomnia, dwarfing my uneventful "literary" life, on a square of floor, between walls of books, looking out onto busses and people

Against Emptiness

even less adventurous than myself, or maybe more, but whom I never managed to know except with my fingertips. The whispering had not come from "the world," in which case I would have acted differently. I would have gone to live in some new city or to work in a factory. Or I would have done nothing at all that I or anyone could see, but would have peeled the film of habit away from the familiar streets, from my friends. The result would have been subtle and private, hardly noticeable even to myself, except that I gradually would have become aware of a subtle reversal affecting all the things I knew: wherever I was, I would, for the first time, have been outside, in the weather.

Instead, in the Sahara, sleeping in a tent, walking at night over a sea of rock, under a cream of stars, huger than anything I had ever known or imagined, the images which came to my mind were images of enclosure; as if I could not grasp this immensity and this simplicity except as a way, not of being outside, "in the world," but of being inside something far different from the world. It was like being inside a soul, I said, or like being inside God, moving wholly within Him. I remembered a shamanistic belief I had read about, and which had, perhaps, influenced my passion for the desert from the very first. On his initiatory journey, the young shaman is often said to be captured by soul-beasts and killed. His body is slit open, and his viscera extracted. The beasts then fill the empty body-sack with rock crystals, and sew it together. When the shaman awakens, he finds that he is in full possession of his spiritual gift, including superhuman mobility and eyesight. He can now move with impunity between the human world and the spirit world. By interiorizing the potency of the mineral realm, the shaman has breached the human limits of habit and mortality. It seemed to me that by digesting the mineral presence of the desert, by attuning myself to it physiologically, so to speak, I would enter into an order of reality possessing neither interior nor exterior. I would not be outside, but further outside.

According to the myth I had constructed, my journey to the desert would be, in its simplest expression, a journey to no place

Three Journeys

at all. It would be an attempt to move into a realm in which the skills of the body would be exactly spiritual skills, and the most practical problems of survival spiritual problems. It was, in my mythologizing, a realm in which the achievements of the spirit would be accomplished by the actions and reactions of the body, not simply by thinking and feeling.

Such is the power of the myth that I continue even now, when I have seen and been confounded by a more complex desert reality; when I have had, by the force of necessity, to integrate "the world" into my experience of the desert, in the form of dusty towns, incompetent mechanics, and careerist government officials; when I have had to draw water from gas station faucets instead of deep wells roofed over with a stone (not once did I have to get out my twenty yards of rope and a bucket); even now, when the barrier of mystery has been seen to lie, not in the legendary dust beyond the *limès*, but in the interpretive powers of my own mind; even now, the myth has begun to reform, as the lips of a wound heal, given time and calm. It has begun to reform, and yet it has changed. It has become more convincing and more seductive than before, because now that I have been to the Sahara and returned, the myth has been clarified by exact images and feelings; it has been enriched by the incredible multiplicity of the desert which drove itself day by day into my experience, feeding the wounded myth, providing me, eventually, with a means to interpret my confusion.

If only I had been able to go further, I have begun to think; if I had been able to leave the main tracks, and take those marked on the map with broken lines, from Tamanrasset to Djanet in the Tassili N'Ajjer, or to Amguid; or the intriguing solitary lines feeling their way across a blank space, marked *piste interdite;* or those, thinner and more solitary, zigging and zagging across the desert like ships tacking in the wind, their necessary course not from departure to destination, but from well to uncertain well, the symbol on the map reading "camel track." If I had been able to travel south from the Tassili, toward the Monts Gauthier, their very existence unknown until fifteen years ago, and then into

Against Emptiness

the never wholly explored Ténéré Desert, at the center of which the map commemorates a bizarre event—the existence of a tree, peeled, craggy, and almost leafless, watered by camel piss; a tree, around which the yearly *azalai*, the salt caravan to Bilma, rallies in mid-crossing: *l'arbre du Ténéré*. If I had been able to approach the most desolate mountains in the world, the Tibesti in Chad and southern Libya, inhabited by a people identified 2,500 years ago by Herodotus as black, swift-footed troglodites, the Toubou, about whom less is known than of any people in Africa. If I had been able to go further, I would surely have crossed that shamanistic horizon integrally, as on this journey I crossed it piecemeal, a moment at a time, so that my journey became a string of interrupted intensities, each hard and bright as a gem, each containing an ephemeral section of the vastness: a camel and rider miniaturized by distance; a rust-colored dune over a black cliff; a line of palm fronds waving crazily in the dimness of a dust storm. For my journey was not a failure.

To be sure, my naiveté fell hard and painfully. Since childhood, I had been disposed to believe in the magic of empty spaces. Natty Bumpo's great plains, Robinson Crusoe's island, or anyone's island at all, provided it was lost and empty, and no one cast away on it talked too much, had been my charmed gate into the marvelous. In my personal mythology, Huckleberry Finn was my friend, Charles Dickens my enemy. Eventually I grew up, which means that I agreed to bury the myth on one of those islands, in a locked chest, so many paces to the left of the great pine tree, and to scuff over the spade marks and water the dust, so that an adult thicket would grow up and hide the spot forever. But one day long afterward, when the thicket had become older and spinier, it began to bear an unexpected fruit: poems about emptiness and abandonment, their hero a shadowy castaway, part Robinson Crusoe, part Huckleberry Finn; an essay about the significance of adventure, drawing its sustenance from the myth which, after all these years, like the dying gods of a more public mythology, had begun its resurrection and escape, rising inside the thicket, coaxing its brittle tips into flower, and then fruit.

Three Journeys

Of all the fruits the most unlikely was this very trip into the desert which had begun several years before as a subject of conversation, but had gradually made itself at home in my thoughts in a way that was new for me. For the idea did not become richer or more complicated with time; it did not attract clusters of other ideas. In some ways, the idea has not progressed in me at all, even now, as if it would not take any part in the rituals of my intellectual life.

For years, the desert became not clearer but denser, more opaque. Instead of orbiting in the lightness of conversation, it unbalanced my system, so that I circled around it, unable to leave it, or to take it, either. I didn't know that its density was feeding on me, on the escaped myth climbing out of my childhood, stubbornly ignorant of the obvious, interested only in its feeling for the magical. The pirate had returned to stomp on the forgotten beach. Huck Finn swooped again along the broad brown river.

As for me, I had never even camped in a state park. I had never hiked anywhere, or wanted to. I had never thought about tents, deserts, sand ladders. My interest in geology had died with primary school. My powers of observation and my interest in observing were small. Until recently, my experience of nature had been mostly literary: I knew what swallows and nightingales did in a poem, but not in the air, and not what they looked like doing it.

My idyll of the outdoors has in it more of Rilke, I imagine, than of Hemingway; more of Wordsworth than of Jack London. I adore the legendary ridicule of the Englishman in the African jungle who dressed for dinner in a black tie and coat. In the Sahara my best moments came toward nightfall, when I had stopped for the day and set up camp. At that hour, the winds became wispy and light. My tent, splendidly green and yellow against the faded color of the desert, offered a contained space which soothed my imagination far more than it protected or warmed my body. A breath of pastoral settled over me, as I unfolded my otherwise useless table and chair, set out my books, brewed my tea (a drink I loathe in normal circumstances), produced my lukewarm olives and equally lukewarm calvados and crackers, and sat back, barricaded behind the red plastic rectangle of the table, forgetting my dried-out hands, my dust-caked face and my light-fatigued eyes.

For an hour I would lounge on the terrace of my mythic Café de la Paix, while the quiet of the Saharan night thickened, and the sunset resolved with astonishing speed into shades of blue, then black, and then stars. It was important, once a day, to be able to put that table between me and the desert, to frame its splendid hostility, and thereby to perceive it, fleetingly but soothingly, from an artistic point of view, which is to say, with distance. Without that distance, I became a bundle of reflexes, a knot of blunderingly developed skills, concentrating on the desert surfaces, on the track, on the coughing and creaking of the car. When I stopped during the day to look at a particularly beautiful formation of rocks, or a scattering of camels, or an oasis, I had to hoist myself brutally out of the slough of my muscular preoccupations, which were still digging out the car, or washing the air filter in gasoline yet once again, and climb onto the high wire, from which it was easy and inevitable to fall.

That's probably why I took so many photographs. It was not to have souvenirs which I would value later on—"later on" was a purely mental construction during the trip—but to accomplish some predictable gesture, some ritual, which would be the physical counterpart of a mental operation. To click the shutter was to see with the eyes of permanence, with the shamanistic eyes I dreamed of, and managed to have only at photographic intervals. The artistic distance which the camera gave me fragmentarily during the day, the table and chair gave me in delicious slow motion during the hour before nightfall.

The confrontation between the desert and the myth was nagging and brutal. The esthetic pitch I tried to coax out of my battered senses emerged all too often with the ominous sigh of a collapsing tire. In the end my spiritual machine broke down more often than my car, which turned out, oddly enough, to be dependable, not frightened as I was by the blackened hulks of cars sticking up out of the sand every few miles, or the engine blocks, pistons, and exploded tires which lay, as if rained from heaven, to mark the transcendental destination of the machine. They had been abandoned in the desert, along with cave drawings, chipped flints, and prehistoric campsites, to signify an event of human willpower, a scar left on the sand by the bizarre breeze of culture. It didn't matter, from the mineral point of view, that they also signified, wreck by wreck, so many human disasters, to which I longed anxiously not to add my own.

Despite its precarious durability, the car came to symbolize the daily collapse of the myth, its ruin in the face of my gift for claustrophobia, which ran wild as the various mechanisms of the car began to fail. Between the all-too-simple reality of the Sahara and my soft skin lay only this fragile clockwork, this heartbeat of cogs and pistons to which I listened all day, diagnosing one imaginary failure after another.

It seems peculiar to me now that I wasn't frightened by the actual dangers of the terrain. Sand and *fech fech* bogged me down often enough. But a shovel, muscles, and patience were all that were needed to get me out again. It might take hours to back out of a sandpit, or to advance a hundred yards, most of it spent on my stomach, half under the wheels, shovelling philosophically in a dusty wind; but that was part of the myth, it exalted me. Even my exhaustion, my lips caked with a salty crust, gave me an impression of strength: in all that dried-out brightness, my fate was in my own hands. But the motor, the transmission, the axles, the suspension, carried me along by an act of grace, it seemed, which could be revoked at any minute. I was merely their passenger, the passenger, therefore, of grace. And it was hard, surrounded by the whine of metal parts, to keep the required faith.

Spiritually speaking, the car was my clay foot. The anxieties it caused me were the glass through which one sees, at best, only darkly. Its very dependability, defying all my bad dreams, became exasperating, because it left me with nervous fantasies, and plenty of annoyance, but no disaster, nothing to feed the myth.

I remember a talk I had with Pierre on the Assekrem Plateau. He experienced the desert as an immense and permanent revelation, he said. It was a language beyond the possibility of human languages, because it was inscribed tectonicly by God. The desert was not dead and desolate, but had passed through the crucible of existence; it existed now on the far slope of life, participating in advance, to some mysterious degree, in the resurrection. But the traveller could only see this from afar, if he saw it at all, because there was always something between him and the desert. When I asked him what he meant, he said: "Your car, for example." I remembered that conversation during the weeks that followed. It was clear that the car was an alien mechanism in the desert, but without it I wouldn't be there at all. My presence depended on it; my presence, therefore, was alien. It was of the desert's nature to propose this enigma.

Unlike the luxurious pantheism of the forest, the desert's vacancy made it necessary to adopt a veil without which a man could not enter into its presence. Perhaps the Arab *chech* and the Touareg *tagelmoust* acknowledge this in some way. The nomad's veil protects him from dust and wind, of course, but it also expresses a sort of humility; it acknowledges the abyss gaping under every footstep, which can be crossed only by the most nimble-footed, the most lean and skilled of leapers.

In Christian Europe one spoke of a *dieu caché*, meaning that the preoccupations of worldly life interposed a veil of seduction mingled with pain behind which God was hidden by the complexity of His creation. But in the desert, the creation has been simplified until a man can survive its emanations only if he takes great care. The saints of the Thebiade believed that a man entering the desert had already stepped beyond the material world, closer to the golden source which glows from behind the saintly figures of Byzantium, dulled and humanized by their faces, but still dangerous, unworldly. The desert was the melted gold itself.

Three Journeys

There it was not God who was veiled, but men who had to be. When Moses came down from the mountain in the desert of Sinai, his body had been roasted by the naked influence of the God he spoke with.

In more ways than one, my car was an Ariadne's thread, its far end tied firmly to my study in New York, to the repetitions and rituals which expressed my need for a safe ground, for without it, the limits of my psyche would collapse. In the desert I felt them weakening, and my anxiety was double edged. On the one hand, it was a weary, familiar pain I could depend on, a sort of negative safety which guaranteed that even here, in the labyrinth without walls, I carried about a ghostly replica of the room I had dreamed rashly of leaving behind me. It was also a reminder that I had failed to get outside into "the world," that my idea of the world had been faulty from the start. My mess of needs and preoccupations, this automobile rocking precariously over the washboard tracks, or swooping through yellow reaches of dust, this very automobile was "the world," or as much of it as I could accept at one time. It was a reminder that, coming into the desert, I had taken my world with me, as a double of worries and longings. This double came between the excessive light of the Sahara and my worldly sensitive eyes, preserving me from the roasting Moses got, but also from the vision he obtained. I teetered every day between vision and worry, as Pierre had already guessed on the Assekrem Plateau, where, year by year, he underwent the initiation by rock crystal.

In the case of Pierre, there was also faith, of which the desert seemed to be an embodiment requiring, as a matter of course, not of discipline, all the elements of the spiritual life: asceticism, detachment from the world, a ripening passion for the One in whose presence one lived by necessity, a melancholy perspective on the Many which, in the desert, was not, in fact, very many.

I don't know much about faith. It requires, I think, more trust and vulnerability than I can manage. If I were to assign a category to my life it would be the category of longing, along with its corollary, nostalgia. But the desert is a sheer presence. Its im-

Against Emptiness

mobility has a quality of nearness which clings. It roasts the traveller who is not provided with a veil, or who has not climbed an interior *chemin de croix*, or who, unlike the Bedouin, is not born into a culture which exalts the insular frailness of life and trains all his faculties to cope with it. In my case, longing became the veil. I managed, in the Sahara, to long for the Sahara; to leap acrobatically into its presence, only to collapse into "the world," and then to leap again, and yet again.

In the Greek myth, Atlas is represented standing on a turtle, carrying the world on his shoulders. According to Ovid, he was metamorphosed into the chain of mountains that forms the northern limit of the Sahara Desert. I went further than Atlas did. I crossed over his mountains, into an awesome country where the Gorgon's dead hair rained down as snakes, and the earth never recovered from Phaeton's escapade with the sun's horses and chariot; where, according to Ovid, the intimate traits of hell were to be perceived. The world I carried on my shoulders was abstract and obsessive: it was my character, my identity. The turtle I stood on had four wheels. I never knew what kept the turtle from foundering, and that worried me a lot. To enter wholly into the desert, I would have had to step off the turtle and let go of the world. Atlas couldn't do it, and neither could I.

PART TWO

Automythology

UNTIL NOT LONG AGO, my experience of the desert had been limited to a child's view of the expanse between the ocean and the boardwalk at Brighton Beach. First memories often have a jewellike quality which later memories are denied. I propose, therefore, this image: a boy, maybe four years old, slipping through a chain fence onto the beach. He trudges in a cool, silvery light along the damp sand near the ocean front. Somehow it is autumn. The eggshell sky, the silvery crust of foam on the waves, and a silvery quality in the sand combine to create a unified impression which the boy will not experience again for thirty-six years until, having scaled a high dune in the Sahara Desert, he encounters the hallucinatory sameness of the *erg* blurring in a sweep of yellow and rust from horizon to horizon. It casts its blur as well into the sky, which momentarily resembles an inverted *erg*, the total view recalling a dream he once had, after reading Edgar Rice Burroughs, of travelling inside the earth in a world which curved up in a vast hollow, so that the sky was simply more earth.

If the boy passes close to or far from any people, they form no part of the memory. It has, in fact, often been difficult for him to include people in his memories. As on this day, people form a boundary which hems the memory in, but they enter into it, when on rare occasions they do, only as ghosts without echo or dimension.

Three Journeys

At a certain point, an adult of great size and kindliness enters the memory. Years later, he would recognize the bulky adult of the memory in the form of Arabian geniis, attentive gods and goddesses and, in general, any kindly manifestation of the dark powers which normally can be expected to act quite otherwise toward a merely human being.

The adult amounts to a dark presence, a shadow, let us say, with nothing to cast it. Later interpolation allows us to pinpoint this dark impression. It is actually dark blue, with silvery buttons; a policeman, whose role is to introduce an entirely new notion into the memory. It is the notion that during all this time the boy has been lost. So many constituents of lostness are absent from the memory that the idea of it cannot have made much sense to him: lost from what, or whom? Retrospectively it is probably important, although even now he is tentative about this, that for several hours his parents have been convinced that he has drowned; that they and several hundred members of a private beach club have been looking for him, or for his dead body; that the police in a whole sector of the city have been alerted.

Revolving around the memory, although at a definite distance from it, is the smaller memory of a room. In the room, the boy's mother is trying to compose her emotions, which are divided between an ecstasy of relief after her morning of morbid fantasies, a nagging sadness at the boy's indifference to what she is feeling, and the obligation, by the book, to invent a punishment which fits the crime. There is no actual memory of the punishment, only of the puzzled expression on his mother's face.

I think I would like to write an autobiography in which there would be no people, although people would keep stumbling into it with an expression of surprise, as if they had stumbled on a shameful scene behind a bush. And yet nothing would be going on behind the bush, the grass wouldn't even be trodden down; no one would be there, or would have ever been there. It would perhaps be the only place in the world of which this would be

Automythology

so. I would write the autobiography of the place behind the bush where no one ever was.

If I am to make sense out of certain choices in my life: out of the emotional hibernation which lasted until I was almost thirty; out of the choice, incongruous to anyone with a view from within of my spiritual existence (but I was the only one with such a view), to become a poet; out of my chance exile from America, which lasted for ten years; out of my addiction to self-awareness, or at least self-preoccupation, the only boy on the block, I suspect, to have gone astray in this particular way; out of the conviction, amounting to a faith, that my life was organized around a core of blandness which shed anonymity upon everything I touched, making of me a sort of reverse King Midas or, more philosophically, a victim of Gyges' ring as Plato never saw him—if I am to make sense out of any of this, above all out of the Sahara desert to which all of it leads, and from which, now, all of it flows, I choose to make the sense begin with this memory. I choose to measure my life not sequentially, as in a story, but circularly, according to its distance from the emanations of this jewel, unspeakably calm, silvery, and vast, exposed upon the colorless velvet of my infancy.

In order to solve an impression of blandness, one can eat spicy foods. If the blandness comes from within, one can fold an element of spice into one's deeds, hoping that the flavor will eventually penetrate to the source of the deeds. One can marinate in one's deeds, so to speak. The effect is a combination of theatricality and sympathetic magic. After a while, the deeds outnumber you, so that astronomically speaking, you begin to revolve around them, instead of them around you.

There is, however, another solution. As you become aware of the creature galloping within like a child's rag doll, giving rise to a boneless quality in the features of the face, a roundness which might be called fleshy if it were not also thin and, even when extremely suntanned, pale. As you become aware of this

Three Journeys

negative creature, native only to the forests of the Wizard of Oz, you can do as the poet Baudelaire, in a similar case, suggested: placing your hands at arm's length upon the creature's shoulders, you can catapult in one naked movement upon its back. Unable to divert its gallop toward the abyss, you can gallop with immense, purposeful joy toward the abyss. The result, in my case, was a form of cowardice also known as being well brought up, well-spoken (in Brooklyn no small task), obedient, except for an occasional episode of hooky-playing.

I propose this other image, more lethal because not personally remembered, but dinned into me by nostalgic aunts, uncles, and parents. A boy in immaculate knickers, his shoes unscuffed, his mind a prehensile organ of obedience to his father's longing for elegance and politeness; a boy so abnormally considerate that a novelist would detect therein a suppressed savagery bordering on cannibalism. The idea was not simply to be good; cowardice alone sufficed for that. It was to be Good; it was to be GOOD. In this, it is said, the boy succeeded, although at great cost. For in order to be GOOD, he had to consent to a feeling of inner monotony, a mood of uninhabited spaciousness, of which his obedience was the climate and the weather. As he was to put it twenty years later, in a phrase which marked the birth of his reflective capacity: "How can I be a writer when I don't have any biography?" This thought still stands in an empty place alongside another thought, roughly contemporary to it, the two forming a sill or threshold, like a pair of fluted columns in a field, the only remains of a never-to-be-restored temple: "I believe I can trace every one of my feelings back to its point of origin in my psyche," spoken undramatically to a friend whose expression of surprise mixed with disbelief I will never forget.

To my friend, it seems, an element of mystery in the psyche went without saying; a murky quality from which dreams and wishes emerged, with an insistence one had no more control over than the weather. This initial premise gave rise to a sum of dispositions known to my friend, and described to me, as the human condition, as one might say a condition of surrender. On this

Automythology

late evening of my twenty-second year, along a rainy street near the Palais de Luxembourg in Paris, between rows of dormant cars, under the periodic glow of streetlamps and a lid of low October clouds, we talked of inner struggle, of unexamined compulsions. of guilt. My friend suggested that anxiety, on occasion, might precede what appears to be its cause; that the self-confidence of one's tastes, one's type of feminine beauty, one's erotic longings, might not be so self-evident after all; in short, that the psyche might lunge along its painful, self-contradictory way with only an occasional reference to that environing principle known to me as "the world." For friendship's sake, combined with a general reluctance to make a fuss, I was willing to entertain these notions. Pursuing my inner gallop (toward O! what abyss?) I was willing hypothetically to allow the view of a human condition in which inner struggle might play a part. Had I not read Freud, Nietzsche, and Marx? Had not a precise language wafted through the mental structure of my college education, studded with scaly words like anxiety and anomie. Anomie, even now, smells to me of chalk and a scarred desk; a voice like a surgeon's voice, placing on this side the ideas, on that the suffering, a division which, at the time, had seemed quite sensible to me.

Although the friend never explained his disbelief, nor did the boy ask him to, it was along a rainy street in the old world, in the vacant solemnity of a discussion of principle, that the boy lost his innocence. It would be months before he noticed any change, and years before the more persistent layers closer to the heart would fall away. There is no doubt, by the way, that we are speaking here of a fall; or rather of a murky spiral around a gray, diminutive silhouette. "In other words, you tend to feel depressed, often." "Depressed, I don't know. It never occurred to me to use that word." This too would be in Paris, although years later. The boy was almost thirty then. Behind him extended an echo on the verge of becoming audible, a structure of glassy tones resembling the cool misty shed of the Gare St. Lazare. This echo had bathed his years in Paris. It had given those years a form, like a tunnel from which he was perpetually emerging, or

Three Journeys

a hall of silences toward the exit of which he had begun to walk more quickly, and would soon begin to run, with the embarrassed hurry of a man on a busy street.

The echo, soon-to-be a whisper, and then even a howl, was the boy's past; was, rather, his discovery that, unlike angels and vegetables he had left behind a trail of selves in painful or joyous postures. They cast a silvery light upon other selves they had known and loved, preserving them and the element of emotion in which they had been sculpted, so that a backward glance into the echo encountered a scene of busy gloom like the one Dante describes on the desert plain before the gates of hell. Every minute of his life, he would soon discover, produced a further length of limbo, an added section of the tunnel from which he emerged as in a dream of frustration (the doorknob perpetually slipping out of his hand, etc.).

"Having a past" had begun years before with the expression on his friend's face. It had been an expression of disbelief, coupled with indulgence, as for a child whose toilet training is imperfect. Gradually the face, with its precocious jowls, its shaving nicks, and its Talmudic gloom, drove a wedge into the boy's obstinate vacancy.

The Parisian spring had blossomed around him, Italy's luminous bulk had swallowed him like a prayer. Each day that summer, on his motorcycle, behind yellow goggles, a guitar strapped across his back and a volume of Proust in his pocket (all of this as if painted onto him in *trompe l'oeil,* simulating a personality), had brought him closer to the edge over which he would tumble a few months later, in an apartment on Canal Street in New York City, to which he had returned briefly in an unsuccessful attempt to begin his life. There would be rats scratching in the walls and the hollow smell of rot, like an olfactory theme, teaching him to play his own tune upon the instrument of time. And he would play, listening compulsively to his heart hollowing out a space each night in his chest into which crept the hairless fauna of insomnia. He played that inner piano, and the neighbors never complained, even late at night, even when he leaned over and banged the

Automythology

keyboard with his elbows. "I believe I can trace every one of my feelings back to its point of origin in my psyche." Had he actually pronounced those words only a few months before? Now they provided a refrain which he set to the music of his dark instrument, like the fol-de-rol of folk songs; a sort of desperate nonsense punctuating the glare of the streets and the nerve-strangling wire of ambulance sirens, and the surrounding plain which was no longer bland, but glossy, mineral, and flat.

II

DOCTORS AGREE that coming into the world is an awful experience which inclines a person to a feeling of pessimism concerning his future prospects. Philosophically speaking, this is probably a good thing, for life could well be defined as the condition of being vulnerable. Rejection, exile, nostalgia, anger, tears, a medley of physical revolutions from sexual orgasm to death: all are inscribed forever in the bared circuits of the organism, so that for every turn of memory a tune blares forth, and for every tune a primeval orchestration, and for every orchestration a return into the gray hills to a cavern covered with hairlike vines from which oracles of paradise emerge, mingled with moans. They form an ancient fabric into which the stem of one's life has been woven, topped by the mystic bloom one set out in search of on that fateful day of red-faced weeping and vulnerability, when one's own hysteria was wedded forever to the terror, pleasure, and fulfillment of a soft, yielding moon, later to be refined into a shape, still later into a form of property: one's mother.

Given the traumatic quality of the initial birth, why would anyone want to do it again? Having acquired the property of vulnerability, why would any sensible person choose to enhance that property in the course of a second, and even a third, fourth, etc. birth, each one repeating the "rejection, exile, nostalgia, anger, tears" which had been so awful in the first place that no one in his right mind can even remember them, though, on occasion, in one's wrong mind, ominous rehashings of the event form an inner weather of dreams, compulsions, and fantasies?

I am speaking here of an important mystery. For there's no denying that one does "do it again," and then, often enough, yet again, in a wearing crescendo which ends, according to religious philosophy, only on the deathbed when, in a blinding clap of the physiological memory, one "does it again" for the last time.

My plan had been to avoid this particular way of making trouble. As opposed to the circular pattern of birth, rebirth, etc., as opposed even to the progressive pattern (corresponding in the political order to "liberal optimism") of an oblique ascending line, I had imagined something like a flat expanse with lots of distance and no direction; a sort of desertlike plain where it didn't matter if one sat down, or walked, or made figure eights, or dug holes. If there was an image associated with this life project it was that of the ocean, but an ocean made of solid material.

The image explains a lot. Its roots clearly are in the boy's childhood, which had been lived wholly on beaches, near oceans. Despite its suggestion of anonymity, therefore, the image preserves a link with his childhood, even betraying an odor of childhood: a mingling of dust and salt breeze. It also recalls the primal gem of ocean, sand, and lostness. Which is to say that its stubborn monotony contains the trick of its own failure, and the key to the boy's eventual fall. For the resilience of his first memory was the blade he would stumble on again and again, until, at the age of thirty-eight, the pangs and contractions of existence expelled him definitively into the scalding reality of the "ocean made solid," which is to say, into the substance of his soul, when he entered the Sahara Desert and had, at last, a biography.

The reader may, by now, have formed an opinion concerning the boy in question. He or she may have remarked, for example, the out-of-date quality, resembling a purity of line, of a character whose innocence, like a chalice in a crowd, had remained intact during the anecdotal stage of his sexual activities (from afar the names, arms, legs, breasts, and faces combine into a classical pose of the sort one visits in a museum), only to start its decline in the course of an intellectual discussion.

The reader may also have remarked what amounts to a reflex of style. What sort of a character, it may be asked, can be foreshortened, grammatically, under the general rubric, "the boy," even when we have swung in associative arcs between his third, his twenty-third, his thirtieth, and his thirty-eighth year? Have we meant to describe a quality of elusiveness and irresponsibility which clings to him, like a Wordsworthian intimation, making him unfit, even at an advanced age, for the sculptural and theatrical dimensions of adulthood? Or are we merely pursuing our promise to circumnavigate, at varying distances, the primal gem of ocean, sand, and lostness, which is shown to exert a pull upon his mature character? Clearly both of these are meant, as well as a third possibility connected to a scene which takes place around his twenty-sixth year.

We are in a small, yellow room, toward the end of the day. The dusk has solidified into granules of gloom hanging over a couple seated half on the floor and half on the bed. Behind one wall the girl's mother is brewing a glum psychological stew. Behind the other two walls, each of her excessively male brothers is performing oedipal operations of a voodoo nature with needles and locks of hair. These influences pour in opaque electric layers into the room, so that the couple must breathe deeply, must even gasp, in order to breathe at all. As if to mock the anguish in the room, the gray-lit bulk of Notre Dame floats out of the night at a distance of magic and low-muttering cars. The girl is sitting on the floor next to scattered paint-brushes and a murky wooden palette. The boy, in several pieces, or so he feels, on the bed, is saying in a strangled, theatrical whisper: "Je ne veux pas être un homme." To make his meaning clearer, which is to say, to lift his anxiety into the intellectual realm, he repeats: "Je ne veux pas 'être un homme,'" hinting at a question of principle which the girl is apparently too obtuse to grasp, because she lets out a moan and begins to cry.

Three Journeys

The boy's decade in Paris, extending with brief interruptions from his twenty-first to his thirty-first year, can be compressed, morally speaking, into the burnished gloom of this scene; more particularly, into the misunderstanding which forms a passionate bond between the two faces in question, one earnest, thin, and puzzled, the other fulsome, blond, and desperate.

III

SEVERAL YEARS before this crucial scene, however, occurred the interlude in New York which had changed his life. For some time, he had been imagining a door frame standing in a field. He thought of an empty lot he had played in as a child: sand, anonymous bushes, a cement blockhouse. He had not walked through the door frame, nor was there any reason to, for it stood alone, without any wall or house connected to it. Yet it appeared that he would step through it one day, and when he did, all his acts and feelings from then on would count, but until then none of them counted.

Stepping from the boat onto the pier, which felt uncomfortably solid and unyielding, he listened distractedly to the voices echoing around him in the tacky wooden shed. He had forgotten that such voices existed, made of the same rough sounds he himself had made most of his life, but now they seemed stifling, vulgar, and dark. For more than a year, he had not heard men yell: "Charley, commeer, gimme a hand wit dis trunk." "What da hell are youse guys doin over der." He felt as if he were being spilled out of a jar. It was a naked feeling, filled with images of the ocean front he grew up on, and the schoolyard where he had hung out for years, wanting to be one of the others playing craps and stickball, and even when he had become one of them, feeling more detached and excluded than before. In the voices of the longshoremen, he heard the exile rising to meet him out of his life.

Automythology

It occurred to him, as he stepped into his father's car and began to cry, that maybe without noticing it he had stepped through the door frame. He didn't know why he was crying. It had something to do with the voices, and the high-roofed shed, and the solid cold floor of the pier.

This was not the last thing he would fail to understand during the coming months until, with mature cunning, he would turn not-understanding into a principle and then—we have come insensibly closer to the northern confine of the Sahara Desert—into a morality. These nine months would be a strange initiation. Although his life would contain many great divides, some towering and abrupt—his divorce—others offering a deceptively packaged air—his decision, a decade later, to return from Paris for good—none would have the lasting effect of this one. Almost two decades later it remains, like the signature of an earthquake, recalling only one scene his eyes ever saw: the cowled pastel summits of the Atakor in the central Sahara. It was, climatologically speaking, as if a new season had been declared, characterized by a new sort of weather, not sun or snow or hail.

Later, he was surprised to recall how little actually happened during those months in New York. He returned from Paris in late August. As if by some trick of echoes, the upheaval which the previous two years had wrought in his existence—he had relinquished his career as an engineer and become a "poet"; violating his personal invisibility, he had learned to play the guitar and sung folksongs in cafés all over Europe, his wariness of the experience outweighed only by his surprise at being able to do such a thing—these startling irruptions had affected his inward being very little, if at all. Even now he carried within him, intact, an image of the most precious lineage, a vase of dusky porcelain from an unknown dynasty, inscribed with a scene which he had never described to anyone, for which the words themselves—undoubtedly ordinary words like shelf, lightbulb and dust—had not existed, but which had guided him safely past the subtle treacheries of adolescence.

I present yet another of the boy's internal gems, a harmonic of

Three Journeys

the primal beach, a theme and variation of the "extended plain," its origin unknown (a dream? a fantasy? a conversation?), although it is probably to be located around his twelfth year. One sees rows of stacked shelves behind a counter. A bare lightbulb hangs over the counter, where several clerks are busy shuffling items across the worn, brown surface to a crowd of jostling customers. The deeper one penetrates among the shelves, the quieter and dimmer they become. Somewhere in the back is a desk, a yellow light, and a person. Only years later did he connect the feeling of sanctity which this image provided with his passion for cloisters. He was the person at the desk. Between him and "people" stretched aisles of stacked silent shelves. The lightbulb hung from a wire over the counter. It comforted the person at the desk to see it across the room and to hear the murmur of voices trickling back to him. His work at the desk was part of that life, was, somehow, a condition for it, but only if he performed it invisibly, with resignation and method.

The image occurred to him, now and then, like a talisman, or a visitation. It was not that he thought of his life in these terms. The image appeared to him more in the nature of a blessing. Whatever he did, the image reminded him, would amount to this. Wherever he was, he would be in this room, busy with the work of bodies and faces, yet removed from them. The knowledge freed him to lead a normal life. He could abandon himself to ordinary acts; he could, if he chose, be "one of the boys." Later he could start to write poetry and travel to Europe. Because everything he did happened under a faint light among cool, dim shelves.

Europe turned out to be as far as he could go and still get back. In his mind, the year abroad was to be a sort of announcement, like the heaps of flat rock one sees alongside desert tracks marking a change in direction, but also, more secretly, providing a rough altar for the god to bless. One imagines the god flowing into the shadowed crevice between the rocks, turning in a slow circle from west to east as the shadow turns, its razory darkness visible for miles over the gravelly brown expanse known as *reg*. Going to Europe was to be the first clear notation in the boy's biography.

But there were perils he had not been prepared for. The apprenticeship of solitude in particular had taken him by surprise. He had responded at first by sleeping all day. Because of it he had felt too vulnerable to leave Paris and travel around the continent. He had needed the small rituals he taught himself to perform: cooking over an alcohol burner at given times of day; reading the English novel *in extenso*, starting with Richardson; walking with exquisite melancholy through the streets and alleys of certain Parisian neighborhoods which came to resemble a weathered gray interior. Above all, there had been the compulsion to learn to speak this language whose impenetrability made him wonder, at times, if there would be anyone to look back at him from the mirror. He had not meant to learn French. He simply began one day, at first methodically, and then with a sort of frenzy, to plunge into the language, as if he might swim through it to the other side of the mirror. It seemed as if another person had sprung loose in the shelter of his being, speaking words which gradually lost their foreignness and became a separate medium, with its own set of dreams and wishes. This, above all, he had not been prepared for. He had, until now, been no more aware of having a language than a fish is aware of having an ocean, and here, suddenly, he had two of them, as well as two poetries, two ways of saying "ouch," and two codes for making love. All of this caused as little self-awareness as possible, for when he felt the tugging of the other self, he became small and still. He felt for the gray shelves, the desk, the yellow light; they were still there. When it was time to leave Europe, he was anxious to close the brackets of the experience. Even before he left Paris, he had already begun remembering it.

For a start, he would enter graduate school. He would climb the "degrees" of a literary career with the same obedient step he had perfected previously as an engineer. Graduate school would not be a vehicle for his ambition; it would be a shelter from ambition. Poet or not, he would gallop more joyously than ever toward the abyss.

After a week he moved into a furnished room on Waverly Place in Greenwich Village which soon was heaped with books, its small space made even smaller by days and nights of patient reading. From his desk he could see the broad tuft of one of the city's nameless trees rising out of the backyard and, like a surreal blossom, the immense stalk of the Empire State Building emerging from the highest branches of the tree. Every horizon was muffled by an enclosure: a room at the top of carpeted steps; a desk in the stacks of a library; a coffeehouse on MacDougal Street, its lights yellowy and dim, its tables faintly sticky. He sat with an old man's stubbornness at the same table in the café, at the same late afternoon hour. With delicious stoicism, he read arguments about the liturgical structure of English poetry. Ghostly vocables studded his notecards, while the angel faces of women strolled past the café window.

He performed these ceremonies long after they had failed him. And they failed almost immediately. One is not to imagine a crack lengthening in a wall. He simply woke up one morning in early fall and looked uncomprehendingly at the books heaped on his desk, at the tree outside his window, at the gray stalk of the Empire State Building. He looked at his nakedness, at his bedclothes. He listened to the air entering and leaving his nostrils. Everything was the same, and yet not. It was as if the elements of his awakening had been taken apart, and then put back together skillfully, but not quite perfectly, for they no longer composed a space.

He continued to read, take notes, and attend classes. He carried home heavy worn books from the library every day, and brought other heavy worn books back to the library every morning. After a month he moved into an apartment on Canal Street which hadn't been lived in for thirty years. The move was a form of surrender, for there could be no "indoors" attached to this apartment at all. Its smell of damp rot, the rat warrens in the walls,

Automythology

the ancient gas heater sizzling ominously in the kitchen all day and, containing them, like an audible frame, the roar of trucks shivering along Canal Street twenty-four hours a day: all of these formed an imprint of his condition which, bemused and almost ironic, he connected now and then to his friend's amazed voice on a Parisian street in the simpler, emptier existence of a few months previous.

The gemlike structure within which he had grown since childhood had blinked suddenly and gone out. He had galloped obediently toward the abyss, and now he was in it, but it was the wrong abyss.

IV

NINE MONTHS LATER, he fled New York for Paris again, not wiser but denser. He noticed now that light had trouble passing through the emptiness of his being. The emptiness itself had become mysteriously cluttered. Shadows of the clutter flapped painfully, unpredictably within him.

Having, despite his lifelong gallop, "done it again," been reborn during nine prickly months; having shivered and wept in the exile of the world, he decided now to exile his exile. Returning to Europe, he shed the self which had betrayed him. It would be fourteen years before he reached the desert. Years during which he would learn to impersonate his existence with the agility of someone who has nothing more to lose.

I think of the human figure in Hieronymus Bosch's landscape of Purgatory, sealed inside a silvery globe. An unearthly succulence surrounds him, but it is oddly without emotion. Its clarity is slightly insane. The artist's detachment and clarity condense to form the globe the figure huddles in. He possesses everything, and also nothing. He has been offered a key to the myth, but the vehicle of his vision is also his prison. Like Bosch's human figure, the boy stepped inside the globe and floated over the insane clarities of a world he could not belong to. He got rid of his language, his nationality. Not yet light enough, he cast out his earlier life, and his friends. This not sufficing, he became their enemy.

One summer day, across a stony valley from the gray-white cliffs of les Baux, in Provence, he entered the world of ideology, of all worlds surely the least likely for this ghostly Orpheus. The girl in whose honor he would one day decide "not to be a man" has asked him the sort of question which previously had made sense to him only in newsprint: "Alors, ces marines, qu'est-ce que tu penses qu'ils font en Liban? Quels intérêts est-ce qu'ils defendent?" The girl was a Communist. This had introduced a delicious feeling of adventure to their affair, which had begun a few weeks earlier, and continued with increasing sullenness, anger, and passion. All of these would go on increasing for six years, especially the former two, until, in a splurge of love and terror, seasoned by his stubborn refusal to take any responsibility whatsoever, and her equally stubborn insistence that he do so, they got married and divorced within a few bracing weeks. But on this prickly hilltop in Provence, the inevitable had only begun to whisper its dark advice.

He looked at the cracked red earth, and then at the sweat trickling along his stomach. It was a hot, blue day. Their clothing lay mingled in a comradely heap. The various uncomfortable positions which their lovemaking had taken, from among the least daring of the *Kama Sutra*, had faded from his joints and muscles. Already she had taken her sketch pad, begun an angular drawing which, to his intrigued eye, seemed neither social nor realist. This would have to be added to her other ideological imperfections: her passion for music (their separation, years later, would be consummated during an organ recital in the Église Saint Severin), her precise knowledge of Gothic and Romanesque architecture (they would make love behind the altars of several country churches of nondescript style), and her ability to cook. In fact she had, from the first, engaged in a number of non-Communist activities, such as shopping, dressing tastefully, and laughing. Most improbable of all (surely the key betrayal, although he did not consider it at the time), she had taken up with him, a relatively satisfied passenger on the ship of capitalist imperialism.

Nonetheless, objectively speaking, what were those marines doing in Lebanon? Whose interests *were* they defending? He couldn't say he knew, and I don't either. I do know that he experienced at that moment a bright flash (a combination of Provençal sun and ideological glare) which, aside from giving him a headache (he was ever after to need sunglasses in bright weather), made of him a Communist. Even after he had put his clothes back on, even after she had put her clothes back on, he remained a Communist.

The time of this scene is 1958, an excellent year for Henry James and T. S. Eliot. The blond dynasty of the New York Yankees is in mid-stride. Gene Kelly is smiling his way through enormously complicated dance steps. President Eisenhower is smiling, too. The Negroes are tap dancing and eating watermelon. There is a difference of opinion among informed individuals as to the meaning of the word "wit." All of these activities are being conducted to the melody of various popular songs, or else to the sounds of Mozart, light and optimistic, who has replaced Beethoven, gloomy and desperate, as music's main man.

But what is the opposite of all these cheerful activities? What lurks around every corner, causing Gene Kelly to smile until his face hurts? What makes most Americans, including the boy in question, want to think about something else as quickly as possible? It is Communism. It is the Cossack horses which André Breton implored to come sip the waters of decadence at the fountains of the Place de la Concorde.

The Communism we speak of must not be confused with a political philosophy. It is more in the nature of a secretion, or a property of the involuntary muscles. Among those medically concerned, there is a dispute as to whether it is acquired or hereditary. It is sometimes represented as attacking the organism from without, and sometimes from within; sometimes as sweeping across continents in the manner of locusts, which proliferate in the impenetrable border valleys of the desert, and sometimes as a moist, low-clinging mist, vehicled by rodents from the cellars of the neighboring houses.

Against Communism, the only reliable protection is a cheerful nature combined with a great deal of ballroom dancing. Less certain, but also of help, is the inner calm which comes from knowing that one has deserved an increase in salary; and, more abstract, of use only in the heat of an argument, patriotism.

It cannot be said that the boy had formed an opinion on the matter, quite the contrary. The very idea of an opinion violated the principle of invisibility to which his life had been devoted, albeit with increasing difficulty and even some recent failures. Indeed, if any claim could be made that he was a man of his time, it would be in this matter of having an opinion. The vacant spin of his inner existence, the suspicious glee with which he had become first Good, and then GOOD, were, it turns out, part of an immense hymn. The rule of silence he had imposed on himself from earliest childhood (one can speak to others, but not to oneself), the thunder of the creature galloping obediently toward the abyss: these were the shelves and the escarpments of the democratic personality. The inner and the outer Good had unexpectedly become one. America danced and smiled, and the boy, with the implicitness of common sense, recognized, writ large, the secret movements of his soul.

It is by no means clear to me even now how such a person ever became a Communist. The logic (or as he would soon say, the dialectic) has not been disclosed to account for such a treason not against his country but against the principle of obedience which had formed his character, his posture, even the musculature of his face. It can definitely be said, by the way, that he did not then nor would he ever look like a Communist, although on occasion he would sound like one, and would observe strictly the interdiction (purely imaginary on his part) against carrying on the decadent activity of a private life.

Failed by the causal presumptions of biography, let us acquiesce

Automymology

in the mystery, and note the dreamlike arc of Hieronymus Bosch's sphere as it sways in the heat wind of a summer's day in Provence. Unable to explain, let us note the coordinates of this day, several miles to the east of Van Gogh's Arles (but how far from his madness?), within view of the reedy flatlands of the Camargues where, at that moment, herds of wild ponies made the ground tremble, and the baroque splendor of flamingos garlanded the mirror surface of the tidal ponds. And beyond these, to the south, a mere five hundred miles away, the crushing monotone of the Sahara Desert, along whose northern edge, sweeping in waves of devastation from the coastal passage of the Nile Delta, across the Cyreniaque and the precincts of ancient Carthage, not halting even there, but destroying and uprooting as they came, ever westward, until the bleak expanse of the Atlantic Ocean arrested their blood-steaming hooves, the Arabian tribe of the Beni Hillal streamed in the eleventh century, their ruthless mobility matching the desert's equally ruthless immobility, speaking its language as the desert saints had once spoken it in another register, the saint and the nomad forming the extreme arc of the desert's myth.

V

SIX YEARS LATER, sunk in a deep chair in the salon of the Abbaye of Royaumont near Paris, he would listen to the asthmatic breathing of an older man whose face he could hardly see. For minutes they had faced each other without talking. It was not his place to speak first, for the man, with cruel irony, had let him know that, as a mere boy, he possessed an attribute which the man loathed but was drawn to nonetheless, as to a vice and a humiliation: it was youth, which the older man, a great Polish novelist, had described provocatively as a sort of original sin which time and pain alone absolved, time and pain being one and the same.

The novelist seemed to be thinking the breath into and out of his lungs, trying to catch the discordant inner music which would mark the final cure of what the other day he had called, with something like self-hatred in his voice, the only antidote for youth.

From the first, he had been fascinated by a leathery, boyish quality in the novelist's face. They had taken long walks together in the park of the Abbaye. They had talked about narcissism and philosophy, above all they had talked about youth. The paradox of the novelist's cynicism was that, in a way, youth had been his only homeland for almost thirty years, during which time he had lived obscurely in South America, his only human connection being to groups of adolescent boys which he seemed to attract, becoming their elder guide and counselor, as he would say, their pied piper to nowhere. His pleasure, he said, had not been to debauch the boys—bleak lines in the man's face made this believable—but to inject an element of vice into their simplest thoughts and feelings, so that even the most ordinary acts would come to seem, and would become, transgressions.

"I will tell you what I think of you," he said, breaking the silence, "and then, when I am finished, you will tell me what you think of me."

Immobilized by the novelist's passionate cynicism, and by a quality which he sensed, although he was far as yet from understanding it, his vulnerability, he listened, as the man began to speak: "My impression, first of all, is that you speak French too well. Even the muscles in your face seem French, and the way you use certain words, 'alienation,' for example, when you mean unhappiness. Yet unhappiness is an ancient, lovely word. It has a patina which comes from many mouths forming themselves around it. God and the devil are enclosed in the word unhappiness. But as a French intellectual you say, 'alienation,' and you feel the march of history at your side. You imagine Karl Marx approving of your ingenuity in finding this new use for a word which was so much more limited in his time.

"I will tell you what I think: this Frenchness of yours is an impersonation caused by fear. You are afraid of being ridiculous. Have you noticed how childish foreigners always seem? When you hear them fumbling for words as I am doing now, or peering from under their eyelids to see how one peels an orange in this country, you can't help wondering if they're not a little stupid. By impersonating a French existence, you conceal your clumsiness from everyone as well as the fact that you feel a little blue most of the time, as if you were looking at people through a glass pane. When a smile or a caress is directed toward you, it stops short by the thickness of a skin, because you're a foreigner. Is it possible that you left America because you were a foreigner even there, and weren't ready to find it out yet?

"You're not so young that this innocence should be permitted you any more, therefore you ought to remember what I'm saying. I am a Polish Catholic as you know. I am also an anti-Semite. You smile, because you don't believe that an intelligent person can be an anti-Semite. Nonetheless, it's true, so you may consider that I'm telling you this through malice; that I'm simply trying to put some scratches on the pane in front of your face. Well, that may be true too. It is hard to see you, because the room is so dark. But even in daylight one doesn't see you very well. If I ignore the impersonation which, by the way, is more artistic than you know, and actually quite unusual, if I disregard it, I see a graceful boy slipping away, but glancing over his shoulder, coquettishly, as if he wanted to be found out.

"In my opinion, you're a wandering Jew, someone who is forbidden to have a home. No power forces him to move on, but the law is applied from within. His existence therefore is bitter. But don't forget, God is a wanderer too. That is why He appears mainly to wanderers, because wanderers exist principally among abstractions. They have given up so much that they have become light and unstable, like winged seedlings never touching the earth."

By the time the novelist stopped, the afterglow of the stained-glass windows had dulled into opaque strips of night. Again the two sat without speaking. During this pause, must we imagine the boy pensive and mute? At the novelist's prompting, if only for a moment, has he glimpsed the vitreous pour of his inner existence? Has he felt the stir of massive roots fishing for moisture in the parched underground, and the vertical pressure of sunlight crushing all movement but that of the wind which gnaws, sucks, and grinds without end? We must not. As yet only nameless hints had reached him of that portion of his destiny which would be compressed into the arc between Saint Anthony and the Beni Hillal, between the Thebiade and the voluminous quiet of the Tanezrouft. To tell the truth, he wasn't thinking at all. He was waiting, and he was intimidated. It embarrassed him that this famous person should consider him, vacant and speechless, not entirely at any given moment a presence, a sufficient subject of interest. It made him doubt ever so slightly the incisiveness of the man's genius. He was, one might say, disappointed.

Happily, the tinkle of the dinner bell enabled him to escape his half of the bargain, for he had no idea how to go about telling the Polish novelist what he thought of him.

With the detachment which characterized so much of his personal thinking, he was aware of how literary their conversation had been. As extraordinary as it may seem, the old man had fashioned the scene in the salon after a scene in one of his own novels. In the novel, however, the words had been more savage. The accents of cynicism and disdain had been sharper. The boy in the novel had been a malleable material, his self-awareness had been cushioned by his smooth and supple body. In the novel, too, he had been bored by the old man's abstractions, but he had also felt sorry for him, as if he had guessed that the novelist's disdain for youth was a form of love, was, in fact, an elegy.

As he walked down the broad wooden staircase and headed for the rectangle of light which marked the open door into the dining room, he was filled with a feeling of exaltation. He began, inexplicably, to giggle, and then to laugh out loud, despite all his efforts to hold back.

Automythology

He turned and walked outside into the park under the bulky shadow of the linden trees. There too he giggled uncontrollably. He felt a mysterious elation, as if a wish he could no longer remember having made had been fulfilled against all expectation, and almost inconveniently. Long after the giggles subsided, a feeling of inner certainty approaching self-confidence remained, combined with an undercurrent of surprise. What surprises most of all in a person who had once made a proclamation of complete inner limpidity was his failure to remember the "wish," so to speak, or to grasp (it kept eluding him) the nature of the fulfillment.

And yet it was simple. It was, one might say, childishly simple. He was simply flattered to the point of giggles at the thought that a novelist might fashion a scene out of his life; that a common measure existed, however fleetingly, between a "character" in a book, and the peculiar bundle of existences which he was. In the dark of his psyche echoed the long-forgotten plaint: "How can I be a writer when I don't have any biography?"

And here, in the old man's novel, even if only at second hand, was a biography. Here was the idea—it was really too much to encompass—that he too, from a certain point of view, might be a "character."

He was now twenty-eight years old. Since his day of headache and revelation in the hills of Provence, he had lived almost without interruption in Paris. He had implanted in himself a complete culture within which he thought and felt, scraped murky bottoms and, more rarely, rose if not to heights at least to silvery and splendid surfaces. In all of this he had become an athlete of invisibility, as one said of the desert saints that they were athletes of God. During these six years he had carried his biography within him, as a river carries the debris of trees, homes, and lives. Among the debris were his Marxism, his Militancy, and his Marriage. They had been borne like uprooted hulks by a current which might have been wind as well as water: the current of his life. But a current is changed by what it carries. It snarls in sunken obstacles, and broadens over settled silt; it divides into separate fingers, making its way through the dark steamy sediment of the delta which it goes on building and rebuilding.

Taken as a whole, this time had provided an answer—or the lack of one—to the outburst of puzzlement and despair which had expelled him from New York years before. His private ideal during these years had been that of a monk in a cell. Although the dusky vase, with its shelves, its desk, its feeling of remoteness, had been shattered, he had set about making another one, and yet another, as an insect, with horrifying patience, rebuilds over and over again the smashed walls of its nest. But he was not an insect. His "impersonation," as the Polish novelist had called it, was more desperate than any theater, because he possessed no other self to be. His "fiction" had allowed him to be in the world, but not of it, a foreigner's privilege; yet his entire being had been vehicled by this fiction. Under its protective shadow, his life had been as active and complete as Saint Anthony's haunted nights. He had been a monk whose cell collapsed about him every day, a nomad (a wandering Jew, after all?) who learned to impersonate the rooted passions of the city so well that his impersonation became a home. Without knowing it, he had made his apprenticeship of the desert in Paris. Among the needs and failures of people he had walked invisibly, like a wind; and yet those needs, those failures, had called from him, in answer, a life for which he might refuse responsibility (that was the quality of "impersonation" the novelist had spoken of), but which stuck to him nonetheless, seaming and sagging his face; which, when he was able to confront it one day, would combine into a destiny.

The Polish novelist had edged him closer to confronting it. His laughter under the linden trees had weighted down Hieronymus Bosch's sphere, and soon the human figure within it would spill into the full flood of the world; his last headlong fall to be across the ocean once again, a decade older than the last time, and almost wiser, which is to say, almost knowing, almost accepting, the odd, self-renewing energy of vulnerability.

Automythology

VI

IF THERE IS a period in the boy's life for which he might compose an elegy, it is contained in the years between Provence and Royaumont; in the "impersonation" which turned those years into a remote yet strangely human fiction.

The figure in Hieronymus Bosch's sphere seems pensive and mournful. His sadness contrasts movingly with the barbarous splendor of the world he glides through. Its dazzling succulence —is it Paradise or Hell?—seems unapproachable, not to him alone, but to us, for its shapes possess too much definition. Nothing of them is left in shadow, nothing blurred by an excess of brightness. Because they exist beyond all change, they are themselves, almost to the point of cruelty.

It is for this reason that heavens hurt the eyes. Not because the god's radiance is blinding, but because even the flowers and grass in heaven possess a burnished overclarity, as if, in their unchangingness, they had been fashioned out of some mineral. One thinks of the intricate decors of Byzantium, or the precise craftsmanship of the eighteenth century, which created landscapes out of metal and moving parts, so that, looking at them through the glass, they seem to breathe and sway with the wind, when it is only a hidden mechanism that moves them. Such heavens hurt the eyes, yet we long for them. We place them on Olympus, or in Blessed Isles. We imagine treacherous journeys to reach them, too dangerous for us to try, but stories are told about those who did, better men than we, who suffered blessings and curses from the gods, whereas we, on the whole, are merely unhappy, or forgetful, and complain more than we should, and are not grateful enough for the glimmers of heaven we get in dreams, in moments of elation, or in the silent skating of love.

Maybe such heavens are not meant to be had at all, but precisely to be longed for. In longing, their mineral clarity is softened by something human, something which comes from us: a blending of fulfillment and elegy, of having (like water after a long thirst) and remoteness (like the liquid flow of the mirage), of self-abandonment and awareness of loss. Maybe this blend is the state of mind we associate with art. When we read a book or contemplate a painting; when, taking a walk, we become esthetically aware of nature's profusion, or of the patina on old buildings or, even more acutely, of objects which a moment before seemed ugly: a subway platform, newspapers blowing against a house; at such moments we are turning our eyes toward heaven and, simultaneously, sharpening our awareness of loss.

What the Polish novelist had offered him that evening at Royaumont (a "biography") was the possibility that one day he might become a poet. From his laughter under the linden trees would come a perception of bright shapes softened by distance, darkened by longing. These would be the images of poems; rather, they would be the medium the images were plunged in, through which they swam and reached their destination, their meaning. His "style" as a writer and as a man would be rooted in a soil composed of his laughter and the pitch black dome of the linden trees.

There is a woman concealed in the chestnut woods, her white skin flashing over wild blackberry thorns and carpets of gray mushrooms known as trumpets-of-death. Could it be the Muse, her robe long since turned to rot, her lovely hair—each strand a poem—filthy and knotted, her fingernails black with mold she has scratched up to eat? Maybe it wasn't here at all, but on a crowded street in the late fall, one year in Paris or New York. The light is almost yellow. A moist wind has driven me into my thoughts, so that I hardly notice where I'm going. Looking up, I see a young girl. She is wearing a drab cloth coat, and a hat pulled down over her forehead. Her face is smooth and thin, almost expressionless, but not quite, as if the outer layer of feeling has been removed, so that, looking at her, one seems to be looking at a naked soul.

Is she the presence poets once invoked in the opening verses

Automythology

of an elegy, according to a precise formula, which was the poet's sacrifice of his selfhood; a prayer for all the voices in the language to pour from his throat in one flawless note?

Suppose it was the Muse? The formula she required has been forgotten. The custom of sacrifice has fallen into disuse. I'll probably never see her again, but if I did, what would I say to her? How could I persuade her to accept my gift: cloth to make a robe, food, love? Will she even know what language I am speaking? Neglect has made her shy and savage.

It seems that I must begin my elegy without her protection. Walking unaccompanied onto the station platform I look as far as I can see into the threadbare darkness of the tunnel lengthening behind me. Through the lens of elegy I see a remote circumference. Within it six years appear as in a sheltering space, seeming more like a life than any part of a life, as if the years had been dreamed, not lived. But when the dreamer woke up, he found that the dream had changed him. It had supplied him with a new language and a philosophy; it had aged and bruised him. The dream disappeared, but the bruises went on healing. The dream-language still came from his mouth, when he opened it to speak. As a result of the dream, he had acquired a past, and could begin, just begin, to write poetry.

VII

NOT THAT POETRY, or even literature, played much of a part in the life of those years. To be sure, he wrote a certain number of poems each month, but writing them seemed more in the nature of an exercise, like watching one's diet. He wrote abstractly, rarely concerned about what the poems meant or about their connection to his emotional life. Writing them at all was an anachronism, for it was the only activity he carried on in the English language. English, the language of his childhood, of his youth, served no other purpose during these years than the alignment of harmonious phrases a few hours a week. It was an impoverished ballet, performed to a music which grew fainter year by year, as the rich discordance of the language, its usages and traditions mingling subtly with his own past, became progressively more distant until the ghostly precision of his poems came to resemble a kind of mathematics.

He lived for a while in a coffin-shaped hotel room on the Rue Mouffetard, l'Hôtel du Midi. Most of the rooms were no bigger than his own, and were occupied by entire families, with children. The poverty didn't shock him because it wasn't accompanied by despair. The people in the hotel led busy lives, full of fighting and laughing and the clink of wine bottles. You could count the floors of the hotel, odor by odor: fried oil, the sizzle of cheap steaks, a faded scent of garbage, all bound together by an ancient woody smell, part rot, part mystery, which pervaded all of Paris' old buildings.

It was not long before he had acquired the habits of the hotel. No noise louder than loud breathing after 10 P.M., although you could scream death threats at the top of your lungs until 9:59, and no one minded. No lights after midnight: a small bulb went on in the concierge's bedroom when anyone's electricity was in use. To disconnect you, she unscrewed the bulb. The concierge must have been a restless sleeper, since one rarely got to read more than a page or two after midnight.

The girl, whose name was Michèle, lived on the other side of the Place de la Contrescarpe, in a family apartment which she shared with several brothers. Over them hovered the iron spirit of their mother, a withdrawn woman who lived in the Alps, she herself an Alp of sorts: steep, cold, and lonely. Yet the mother possessed a wild youthfulness which, even at a distance, could be overpowering, all the more so because it was inarticulate. Mostly it took the form of tenacity, a childish, head-on quality that could be disguised as timidity or innocence, but shed a dogged light on everything she did, especially on her children.

All three possessed treadmarks where the mother in her innocence had rolled over them. Into the treadmarks seeped anger, and a primitive desperateness which they called love. The four together formed a sort of primitive horde. The father, a Communist, had been killed during the Resistance, not by the Germans, it was rumored, but by Gaullist Frenchmen, and the bitterness of his death had not been allowed to fade. As a result, the family's Communism had a mysterious, almost ritual quality. The mother was the priestess, emanating a spirit of militancy which had become exasperating to the children because, in comparison, their own devotion seemed full of flaws, being merely human. Worshipping their father's memory, love in the horde had become an abstract emotion which hung loosely and clumsily when applied to actual people, but gleamed fiercely when aimed at an ideal.

Toward each other they were devoted and unforgiving, a combination which tended toward permanent struggle. The savage dance they performed around the mother was fascinating to observe. They attacked or cajoled her, they wept with tenderness or rage, and went out regularly to skirmish with the powers of social injustice, under the gaze of their father observing them in Dantean profile from a blurred photograph in the living room.

From the first, the family, in its faded apartment overlooking Paris from the summit of the Montagne Sainte Genevieve, had been a challenge to the boy. Apparently these people had never heard of Gyges or his principle. They had no respect for his invisibility, taking it simply for cowardice or, worse than that, dullness. When he performed his inner ceremony, when he glued the wrecked vase back into place, the dust, the shelves, and the remoteness emptying the expression from his face for a soothing moment, they spit on the floor and broke some furniture; they belched an insult, and went out to eat some fascists.

His own Communism had thickened, if not deepened, during the autumn months, although compared to Michèle's blond, mystic passion, it would always seem puny to him. He quickly recognized his inferiority, and accepted it. How could it be otherwise? His genes had not been fused in the crucible of the Resistance. His father had not been shot by traitors. During the war, he had merely gone on being a child. He had learned how to read, and gone swimming. He remembered a failed victory garden, and some minutes spent underneath his desk at school during air raid practice. But try as he might, the war reduced itself in his memory to a series of maps in the newspaper, in particular a pendulous black line that changed shape every day and was called the Belgian Bulge.

He did his best to make up for this. He shrugged America from his shoulders almost overnight. With a mental scrub he cleansed himself of the imperialist cobwebs which had falsely humanized his vision. On the pyramid of the global class structure he had been born high and narrow, so he had a lot to expiate.

Wielding the sword of judgment, he read a barbarous cocktail of newspapers. In the morning, *Humanité* and the *Herald Tribune;* in the afternoon, *Le Monde;* on weekends *Humanité Dimanche* and *The Economist;* on odd weekdays, one or another "spineless" radical weekly. Plus the entire leaden gamut of Party publications. The purpose of all this reading was not to find out what was happening in the world, a pleasure which would have smacked of English pragmatism. It was to identify the secret script sprouting darkly, inexorably from the contingency of what was merely happening. In his rather primitive view, the world had become a network of signatures and codes which one had to winnow from the chaff of events and interpret according to the voluminous *clef des songes* written collectively by Marx-Engels-Lenin-Stalin.

To tell the truth, the boy never actually enjoyed reading *l'Humanité*. No matter how hard he tried to turn the judgment against himself, deploring the bourgeois estheticism he had apparently been condemned to from birth, the newspaper seemed too strident, too simple-minded, and too full of grotesque clichés to read.

In times of stress, the individual, it has been remarked, may reach into the sealed well of archetypes and refashion for his private use some age-old ceremony of the race. In his dilemma, the boy resorted to an ancient oriental practice, the most prestig-

Automythology

ious example of which is the Tibetan prayer-mill. One takes a good-sized barrel and fills it with holy texts. One sets the barrel on a revolving axis connected by a series of cogs and belts to a water wheel. The turning of the wheel communicates a rotary movement to the axis causing the barrel to turn upon itself. Between the One and the Many, an agreement is reached that a single rotation of the barrel causes every prayer in every holy text within the barrel to be said once. The boy's Western background, combined with his brief experience of engineering, enabled him to simplify this procedure. He simply bought *l'Humanité* each morning and put it in his pocket. In the evening he took it out again and threw it away. The act of loyalty announced by the title peeping from his pocket; the modicum of bravery it required to saunter past a police station with the telltale label prominently displayed: these constituted, in the archetypal mode, an act of reading.

During these months, which became a year and then two, he experienced what I can only describe as an immense, unspoken hope. This hope, on the epic scale, transformed his sense of the people he had begun to meet. Removing them from the scuffed world, it placed them in a dawn of invisible perfections which was almost palpable, almost whitening: the dawn of the Revolution. In my opinion, he did not simply imagine this. Even in the late 1950s, when it had become a political anachronism, the Revolution lived on in the gestures, tones of voice, and precise mental frame of great masses of Frenchmen, not, strictly speaking, as a political objective but as an aura suffusing their lives, as if every moment, no matter how ordinary, were lived on a rooftop in expectation of the Seventh Day, tomorrow, which would release not only them but all of creation from its bondage to mediocrity and suffering. It was all the more dazzling that these people were not intellectuals, but shopkeepers and factory workers.

The strongest, although almost whispered, voice among them belonged to the leader of Michèle's cell, an old trade union militant named Jean Dejante. Decades of struggle had been worked into his body: scars of police clubs; the broken, badly set bones which caused him to limp slightly; an out-of-breath quality in his kindly voice, the result of tuberculosis or some other ailment caught in a prison camp during the war. Among the cell members, only he seems never to have entertained the notion that the boy might be an American spy, or that Michèle, by living with him, had taken a first step into the gloom of heresy.

Dejante's wife, Riri, suffused the tiny shop they lived over with a dazzle of white hair and ample folds of body. She spoke a giggly sort of French, well described by her nickname, and seemed to the boy to embody pure hope. She sat all day, enthroned in the window of the shop, knitting the socks of the Revolution, speaking with the broad earthy accent of social justice. That she was also a relentless gossip and a promotor of the spy theory never bothered him. Although he wasn't a spy, it seemed to him that he deserved to be thought one. Indeed, from the perspective of the global class structure, it was only by a miracle that he wasn't one. As for the notion that he might be a bad influence on Michèle, he dismissed it. She was quite obviously beyond the influence of such as he; that, indeed, being the hidden comfort of their relationship.

So it was that the idea of social justice took up its orbit in his personal mythology. He ate its crusty bread, and soared into its sphere to the accents of "l'Internationale." It released him, and would release the world, from the failed rooms and broken vases of a merely private life.

In condensed order, a century of clichés bubbled past his lips, refined from the crude matter of *l'Humanité*, but strictly speaking, of the same moral-chemical structure. He existed in a circle of abstractions, closed off by such then-clear now-mysterious notions as *materialism, idealism, historical necessity, objective analysis, internal contradictions* of which capitalism apparently perished, *dialectical* as opposed to *formal* logic. These were not simply words or mental tools; they were tropical certainties, the efficacy of which was so great that merely to pronounce them relieved him of all anxiety for the day.

Social justice was not only desirable, it was a radical treatment for the human condition, like one of those patent medicines which cure tuberculosis, bunions, and the common cold, and are also good for polishing your shoes. Social justice would solve the burden of privacy, loneliness, and personal emotion. The idea of sexual impotence would cease to terrorize him. Instead of the friends he longed for and didn't have, there would be a race of brothers and sisters, foreshadowed by the ritual bond contained in the word *Comrade,* which made him blink and want to weep. At times it seemed to him that a world full of terror and clichés was nothing beside the warmth he felt when someone he didn't know, a garagist in a small town, or a fierce but careful peasant on a country road in Castille, called him "Comrade."

For a time he experienced the idyll of a totally justified existence. With a feeling of inner luxury he stood among the brave faces vending *Humanité Dimanche* on street corners, ready to scatter defiantly at the provocation of anal-featured policemen, or to advance, brother and sister alike, against the assaults of political adversaries, generally referred to as "fascists." He sang American work songs, accompanying himself on the guitar, at social occasions organized by the local Party section. He consumed gloomy blocks of theory, reading Marx, Engels, and Lenin in the spidery print of Moscow editions. He engaged in ruthless conversations, the trend of which was to deplore the dimly outlined atrocities divulged by the XXth Party Congress, but which could scarcely have been avoided, it was sadly agreed, in the face of world capitalist viciousness.

There were winter evenings and afternoons spent drinking tea on the floor of the apartment with Michèle's brothers and their friends, the humorless fury of the conversation matched only by the cold wind stabbing under loose windowpanes, causing them all to hunch deeper into their sweaters. The more they talked, the further to the left the political line shifted, beyond which only fascists prowled in wealthy abjection. From the ideological plush of that room was to emerge, only a year or two later, one genuine terrorist: a gentle, almost effeminate boy, who didn't say very much, and often brought along photographs he had taken displaying a profound, nonpolitical tenderness for existence. He was drafted into the French army, underwent a reversal of signs in the mixing bowl of the Algerian war, and came back with a briefcase filled with excellent photographs of the Algerian peasants he had killed in the course of his masculine "acting out" phase.

VIII

AS TIME PASSED in the radical nest on the Montagne Sainte Genevieve, the political discussion did not result in much political activity, the only real "action" coming early in the boy's conversion, at a national rally convoked by General de Gaulle on the Place de la Republique, in the fall of 1958. The Communist Party had called for a counter-rally at the same place and time, and he and Michèle, among other neighborhood militants, had offered their bodies to be converted temporarily into masses, as the French saying—*faire masse*—with rare cynicism, goes.

As they came out of the Métro into the steaming presence of the demonstration, he felt awestruck and hushed. Here were men in overalls, thousands of them, pouring into the center from the mysterious northern suburbs where the working classes brewed the fire and flavor of Revolution. Their faces were rough, self-confident; cigarettes hung from their mouths. In the solemn shuffle of the general will, they cracked little jokes, and laughed, not raucously, but with purpose and restraint.

The crowd thickened as it flowed from side streets into the Avenue du Temple where, as far as one could see, under red-painted banderoles, like the enormous quiver of one powerful muscle, the People waited, shouted, sang the "Internationale" with atonal thunder. The boy was transfixed by the dignity and power of the demonstration. To him it seemed as if an invisible nation had arisen within the nation, a society of sisters and comrades, fused into a rough whole by the passion for justice, and the pained awareness of injustice, above all by a shared belief in the Seventh Day, tomorrow; as if this merged, angelic mass had come together to be visible and, by its presence, to judge the elegant charade of the other nation, dressed in silver and silk, which had never been young, which rode in black cars, and spoke with the toothy lilt of sexlessness and social privilege.

The podium was lost in the blue distance of the square, one never learned precisely in what direction. For whatever lofty message the General planned to extend over those massed, expectant faces, waving his touchingly clumsy arms and falcon's nose, not one word or gesture reached the outlying space of the Avenue du Temple, where crowds of citizen soldiers thundered their answer, their seamed faces shrugging off the mental burden of oppression, by an act of collective rage and understanding.

He stood in the crowd, shouting the few stanzas of the "Internationale" he could remember (he never quite mastered "The Star Spangled Banner" either); he picked up the slogans that rumbled by him like a magnificent verbal weather. In short, he was in a state of exaltation far beyond anything he had ever experienced, even at the most exalted college football game.

Then, without warning, half a dozen black police buses knifed into the crowd one behind the other, to form a barrier separating the Place de la Republique from the Avenue du Temple. He found this impolite, and was about to say so, when he heard a distant, feathery rumble, as if the ground had filled with subway cars. One couldn't tell where the noise was coming from. For a moment, hushed voices combined with a faintly hunted expression in people's faces to create a tremulous stillness, which collapsed all at once into screams and shouts. A bulge of bodies pushed all in the same direction. The feathery rumble was the sound of police clubs thudding against the bodies of demonstrators. And it wasn't so distant any more. People with blood on their faces and arms dangling down like ropes had begun to stumble past.

He grabbed Michèle's hand and ran blindly with the crowd, until they were brought up short by a tangle of bicycles and a wall. One of his shins had been torn by a loose bicycle peddle. They could see the black leather uniforms and robot helmets of the police coming toward them, and the dark streaks of rising and falling clubs beating at the demonstrators.

Noticing that they had been squeezed against a door, Michèle felt behind her for the latch. It clicked open, and half a dozen people fell with them into an empty hallway. Almost immediately someone pushed the door shut again. Now they crouched in the pitch blackness at the back end of the hallway, and waited. If even one policeman had seen them tumble into the hallway, they were finished. Any second now, the door would be torn open. A few screaming cops would pour in, and mash them into the space under the stairwell, fracturing skulls, breaking limbs, and rupturing spleens. They breathed the moisture and stink of the closed space, and they waited.

After a while, the thudding outside stopped. Someone felt his way to the door, opened it a slit and, without saying anything, stepped outside. One by one, the rest did the same. After hesitating for a moment, he peeked through the partly opened door; then he too stepped outside into a scene of utter strangeness. Empty shoes lay everywhere, imparting an air of startled abandonment to the cleared space. Policemen walked around talking quietly among themselves. Their bat-sized clubs dangled from their belts. They smoked cigarettes and laughed. Nor did they seem to notice when a young couple tiptoed past them. It seemed to the boy as if they were wading through an iridescent liquid that might crystallize at any moment and crush them forever.

IX

AFTER that heroic climax in 1958 diverse pressures began to disturb the boy's equilibrium on the pinnacle of "Stalinism," as he was soon to call it. A historical period was coming to an end in

Automythology

France. At the very time he was breathing sighs of abstract fellowhood, and sheltering his merely private life inside the Boschian globe of world Revolution, the globe had begun to wobble and dent, spoiling its most attractive quality: its heavenly surplus of clarity. As long as the Seventh Day had loomed without question, people had been certain of who they were, and why. Introspective pessimism had merged, at the highest level, into the collective face on the Party poster, jutting its jaw into the sunrise. But a turning point had been reached during the 1950s, no one seemed to remember when: a snag awash in the sea of historic events had rammed the collective vessel, making it shudder a little, and tilt. Now, one by one, in the vessel's sleepy depths, passengers woke up, sweating, trembling for their lives, for it had come to them as in a nightmare that the ship was going down; that the Seventh Day whitening on the rim of the world had been put off by the ineptitude of history, had been, maybe, cancelled.

Here and there in leftist circles appeared men who blinked and rubbed their eyes; who looked around them, a little frightened, as if wanting to be told what to do next, before remembering with a start that from now on they were going to have to tell themselves what to do, which had been precisely the point. Men who came of age during the Resistance; who had perhaps risked their lives to rub out the shame of German occupation; who after the war had devoted every sinew of personal ambition to the collective thrust battering at the door of the future, not noticing, during this time of heroic obedience, that they were growing older, that the dialect of clichés they had accepted to speak had become glued to their identities, had ceased to be a mask humbly accepted and become instead the fiber and substance of their faces. Men in their thirties, forties, and fifties whose marriages, professions, and friends had been experienced as rivets in the inexorable machine of Revolution. Except that, one day, they had stopped believing. Maybe it was Hungary, or the XXth Congress; maybe the minor intricacies of their existence under Capitalism, the tender labyrinth of daily life, began little by little to corrode their principles, trapping them into a love of life in the present world, however imperfect, and an ever-so-slight boredom with the Future, however full of songs.

Each of these men bore the marks of a terrible awakening. Some had quit the Party on their own, others had been expelled. But each had heard the gate click shut upon his past life. In the eyes of many they had done more than make a decision; they had wrestled with the angel. They had accepted having their beliefs, their structure of moral comfort, their formal link with "humanity" across the abyss of class and taste, thrown down as by an earthquake. And now they wandered among the ruins like Job's servant, whispering: "I only have escaped alone to tell thee." It was a magnificent whispering, tragic and brave, or so it seemed to the boy, who listened.

They sang solos expressing the anguish of their political rebirth, the confused but youthful energy which resulted from being free to entertain ideas for their own sake. They wrote of Heidegger and Nietzsche, of linguistic theory, of group dynamics; they analyzed the "mythologies" of daily life; they discovered that the "superstructure" was not simply a disease upon the basalt of economic reality, but a network of subtle pleasures, a penumbra of relationships which evolved according to laws more elusive and complex than any "Marxist" theory had attempted to account for.

In their quest for a new Master, they did not abandon Marx, they regressed him. They discovered the "young Marx," the Promethean moralist who, in his youth, it seems, had made humanistic notebook jottings which authorized their apostasy from the bulky dolmens of Communist theory.

Looking back on the excitement of the period, it appears that these men devoted a great deal of exuberance and showmanship to discovering what, on the whole, many people had already known for a long time, and that the boy was getting his education under somewhat false pretences. Ideas on all of these subjects were probably being noted down at that very moment in neat, book-lined studies all over the world.

Automythology

A few years later he would come now and then across some new work by one of the great figures in a bookstore. He would pick it up, and for an instant the febrile excitement of those months would tingle in his fingertips. He would remember the flushed faces of the debaters in Left Bank lecture halls speaking words full of inner whispers, like *alienation*, invoking that rough Beethoven of social philosophy, *le Jeune Marx*. As they stood and gestured on the platform, they had about them the craggy glow of the pre-Socratic philosophers who became immortal not simply because of their ideas but because they invented the medium of thought itself. He would remember the smell of mildew and unwashed feet, the edge of chill in the air, the bulky sweaters, the raincoats folded on laps over piles of books. Mountains of theory were displaced in those faded halls under fly-specked lightbulbs. And these men were the apostles. One with his mane of white hair and equine face, sculpting his eloquence with spiral movements of his hands; another, his usual adversary, clutching the table in front of him, his voice passionate and misunderstood, his folds of fat and his almost trembling immobility resembling some vast armored creature; still another, shrill and boyish, with round eyeglasses and a flustered manner.

All of this would course through his fingertips. He would open the book, and he would find an excellent treatise on the peasantry of southwestern France, a cogent exposé of the problems of modern urbanism, or yet another interpretation of yet another text in the Marxist canon. The heroic moment had come, and it had gone. For a short time, a few men had been inflated beyond their personal limitations. They had dreamed aloud of a unitary discipline in which all the branches of the human sciences would be combined under the auspices of the deepest personal emotion, and the nobility of literary style. They had offered works of confession which were also instruments of thought, as if through colonnades of Spinoza there had blown a breeze of Kierkegaard and a wind of Nietzsche, inscribed, to be sure, in a general climate of Marx.

But by the early 1960s, the heroic flood was over. Once again sociologists and philosophers, psychologists and historians went about constructing works of drier, more solid material, meant to last longer and to resist bad weather, but not to exalt the intellectual passions of a young man in whom thoughts and ambitions had been implanted for the first time.

It is, of course, possible that he simply misinterpreted the excitement of the time. The tale of those charged and glowing months may merely have been his own tale. But the model he received was powerful and real. It would mature for a decade in the slow glue of his mental processes. From then on, he would measure his ideas on an epic scale. He would dream, with much groping and failure, of an intellectual medium which would also be a passionate medium. He also, by no means incidentally, ceased to be a Communist.

For a while he transitted in a sort of Revolutionary Protestantism, which had turned Paris into a nursery for "splinter-groups," "study groups," "anti-groups," and "groupuscules," orbiting around the Communist Party like gnats trying to sting an elephant, but mostly stinging themselves. His principal weakness as an activist in these groups was the trouble he had staying awake at meetings. It had been easier to crank the shaft of the prayer-mill than it was now to witness the sinking of the French language into a limbo of abstract substantives. *Liberation, materialism, dialectique, alienation, praxis* and *practico-inert, dédramatization,* and *désalienation* circulated like money subject to permanent inflation, so that ever larger quantities of it were needed for the most ordinary transactions. Finally it was replaced by scrip; in his case, by sleep.

X

BY THIS TIME, he had moved from his coffin-shaped hotel room into the family's apartment on the Rue de Cardinal Lemoine, where the exquisite pastoral of the city glistening through the

Automythology

window contrasted with all the fulsome varieties of anguish within.

The extended family included the brothers, each with spouse, he and Michèle, and on occasion the mother visiting from the Alps. In addition there were several friends of the younger brother: a depressed dental student; the slightly effeminate photographer who, everyone knew, was in love with the younger brother, although he finally settled for an exhibitionistic affair with the brother's wife.

One room of the apartment was devoted mostly to acting lessons and hysterical shouting—the first brother; another room, to exremely complicated sexual permutations among wives and friends—the second brother. The room he and Michèle lived in stood for hard work, ideological purity, and discretion: they shouted and wept only when everyone else was out.

Later he thought he recognized this room in certain Renaissance paintings of Saint Jerome: a small, massive desk piled with books; beside it, an easel holding a murky unfinished painting; at the foot of the easel a palette thickly sculpted with dried paint; next to the easel an overly narrow bed which had turned the couple into nocturnal acrobats. The space of the room was pulled together by an irregular patch of floor covered with a rug, the intense clutter of the whole framed by a window overlooking half of Paris. It was the monkish aspect of the room, its atmosphere of compressed meditation, that made the boy think of Saint Jerome. In the window's broad frame he saw a brooding figure bent over a book, the face expressionless, the body endowed with a quality of stillness best described as invisibility. What is the left hand doing in a corner of the desk? Could it be fingering a death's head, as in George de la Tour's famous painting of Marie Madelaine? Surely not. Yet at moments, the boy seems almost to include this act in his memory of the room.

Here the summation to "be a man" had been boldly refused. Here, all the acts of a gladiatorial relationship had been performed year after year: his sexual failures and his more numerous but somehow less visible successes; the fabulous balancing act which the couple had worked out between them in the space normally occupied by tenderness and love; the body-throws, and more rending, the soul-throws. On the whole, he had been too busy keeping on his feet to enjoy the luxury of self-knowledge. The heroic dimension of their relationship had formed a bulwark against the most unsettling challenge of all: the challenge of tenderness, to which neither of them were equal, although an important difference existed between them on this score. He, on the whole, was satisfied with the arrangement: it preserved within him a vacancy he did not wish to have disturbed; she was not: it caused her an anxiety which expanded gradually into a nightmare, the nightmare that, for all her splendor and passion, for all her genuine beauty, she was not lovable.

Year after year she hurled herself against his impassibility without success. Maybe she would have done better to charm her way into the dusky cell which he held desperately about him like an outsized pair of pants. Maybe, but then again, maybe not. For he might simply have run, and kept on running, until he reached Tangier or Rome, shedding ideology as he went. He might have gone on retreating into the coppery flatland of the Sahara Desert rising ever so faintly toward the horizon, simplified by the inner hum of salt percolating within it after each rainfall, so as to form at the surface a blank crusty soil: the blazing *sebha*, reflecting back to heaven its lack of a face and of a body.

But it was not time yet for Parthian victories. Another lady, in another year, would grind the stone of panic into his meat, before casting him aside, his invisibility lying about him in ruins. The importance of such defeats from an archeological point of view was as yet unknown to him. It was not the breaking, but the desperate making which concerned him now, and Michèle was his accomplice, to her cost. For in their intimate division, the fear was hers, the armor his; a truth which he glimpsed only on rare occasions, and then usually with bewilderment, as on one intensely blue day in Arezzo, their eyes still dilated by the hieratical vacancy of Piero della Francesca's magnificent frescos.

As they walked through the cobbled streets of the old town, they came upon a church with its massive doors swung open. From their place in the bright July sunlight, the interior seemed pearly and cool. They climbed the broad stone steps and entered the Church looming emptily before them. In the central aisle, in front of the altar, stood a narrow table draped in black and marked with heavy gold lettering. When they approached, they saw that it wasn't a table after all, but a coffin. Patches of bright pastel were strewn over the coffin, shed by the stained-glass windows. He didn't feel any emotion, but the visual pungency of the scene penetrated him like a high C: the abyss-colored coffin, flaming spots on the floor, rows of empty pews bathed in cool and shadow.

These perceptions were fleeting, however, for Michèle gave a low scream and ran out of the church, leaving him alone to his esthetic detachment which made him feel inadequate and foolish suddenly. Her morbid terror was out of keeping with the Mediterranean sunlight and the flowered stone terraces of the town, yet he wondered if it were not, in some respects, superior to his appreciative quiet.

He had once thought of a man walking over a thin layer of ice. Beneath the ice the water was freezing and deep. The man slid his feet gingerly along the ice, as if, at any moment he might break through. The image, for him, had had a philosophical dimension: it signified the precarious security of surfaces, as opposed to the deadliness of depths, in a world generally characterized by cold weather. The image occurred to him again as he walked through the empty nave and out into the street. For it was clear to him that Michèle had broken through, that she was struggling in the chill of useless profundities on the raised curb of the sidewalk across the street, where she sat, her shoulders hunched forward and her face buried in her hands.

At such moments he had a tendency to pull back—the expression in his mind was "hang around"—for this was a vein he preferred not to encourage. Anger, indifference, competitiveness, yes; but not this. Not these abysses of need and vulnerability; not this mute plea for something beyond their usual companionship, however incandescent. As on several other occasions, he put his arm around her shoulder and sat beside her uncomfortably—still very much "hanging around"—until it was over. Several years later she would accuse him of too much "side by side" and not enough of all the other moral-emotional positions. Even in the midst of his colossal anguish at their separation, and her departure for another man, he would be puzzled, for "side by side" had seemed fine to him, and brave.

XI

IT IS POSSIBLE that these "internal contradictions" would have brought the couple to grief far sooner than they did were it not for the Algerian war which overran their personal lives suddenly and completely. During the preceding year or two, the war had gradually become an environment pervading every aspect of their existence. Broken heads and limbs, jabs, gouges, and bullet holes were a casual part of the landscape. As a matter of course, people crossed the street before coming to a police station: within those deceptively open doors a mad regime was known to exist, its law extending democratically to anyone passing through them. This law prescribed beatings with fists, clubs, feet; imprisonment for days without the bothersome paper work of arrest and accusation; sudden and apparently mortal illnesses.

There were interrogation centers in outlying suburbs, where Arabs were fire-hosed, starved, and otherwise "treated" for the fever of hope which the war had injected into their lives. In retaliation, squads of Algerian rebels—their organization was called the F.L.N.—went around shooting at police stations, police cars, and lone cops. They also shot at each other, applied knives—known mysteriously in French as the white weapon—to traitors in their ranks, and used death normally as a means of doctrinal persuasion.

The atmosphere was of terror and vulnerability. One day in the Rue Mouffetard he heard an echoless bang in the crowd up ahead of him, and saw a man walk calmly down a side street, leaving behind him a stunned silence and a body collapsed on the pavement with a bullet-hole in its head. Later he remembered feeling more bewildered than scared. For a moment, he couldn't remember why people talked to each other at all; why they kissed, touched, and had kind thoughts. Why, in short, they didn't kill each other right away. Acts of love seemed so improbable, like specks of rock in the middle of an ocean. In his bewilderment, it was not the murder, so casual, so unworthy of special mention, which surprised him, but the fact that it didn't happen all the time; that every conversation didn't culminate in a murder; that he himself had somehow never killed anyone.

The nihilism of his thought weakened the solidity of such words as *imperialism, colonialism,* and *world economic oppression,* for it confused his desire to know who the enemy was and who the ally. At times he experienced a change in his thoughts which almost sickened him with its suddenness and its possibility for despair. His dream of brothers and comrades, conceived with such fierce moral energy, became brutally inverted; it was no longer an irresistible music flowing over the earth, but a desperate raft teetering on a mad sea, a raft that was foundering.

During one of these moments he recalled a phrase he had once read in a book about the inner feeling of schizophrenia. The patient, with the pungency of an oracle, had said: *"La terre bouge, elle ne m'inspire aucune confiance."* ("The earth moves about, I can have no confidence in it".) These words conveyed a naked feeling which he would experience again, twelve years later, with a thrill of recognition, on the plateau of the Tademait, north of In Salah. Sullen and brown, the immobility of the desert had seemed to him a form of moving about, perhaps the most profound; a steady motionless tremor which shone from within the human vision, shaking it to the verge of terror.

From the start, the war had affected him personally. The Rue Mouffetard, with its string of cafés and shabby residence hotels, harbored a large Arab population which stood quietly all day in front of the cafés. The Arabs kept their backs turned to the European world. They rarely walked alone, and never at night, when a policeman's whimsy might turn into cripples or corpses.

One day a story reached the small café at the top of the street where he took his early morning coffee. An Algerian had been found seriously wounded the night before on the Rue Lhomond. When the police were informed, they posted a man across the street to keep watch while the Algerian died. It took all night. They didn't call an ambulance or a doctor, they just waited. The rumor was that he had been shot from a passing squad car—a form of demented target practice that was not unknown in those days—and the police were in no hurry to have a civilian doctor at Hotel Dieu extract a bullet of well-known caliber from the Algerian's wound.

The violence of the war wheeled around them until, quite abruptly, it submerged their lives. A friend of Michèle's came to see them one day. She wanted to know if they would be willing to give some form of personal help to the Algerian rebels. It would most probably involve transporting messages or packages of money collected by the F.L.N. from Algerians working in France, to finance the war; they might also be asked to provide a safe hideout for F.L.N. guerillas, if needed. He knew that such clandestine groups existed. The most famous was the *reseau Jeanson* organized by a friend of Sartre's. But there were others too, whose declarations and arrests were periodically reported in the newspapers. The prison at Fresnes had become a revolutionary academy where political prisoners studied Marx and Franz Fanon, gave courses in revolutionary theory, and debated each other under the indolent eye of the prison guards. The heroic time of Lenin and Trotsky had returned. He was stirred to learn that Trotsky had actually occupied a room on the Rue de l'Estrapade,

Automythology

and that Lenin had slept on his floor. Or was it the other way around? It turned out, too, that Pascal had lived behind a window which he could see when he wrote at his massive desk. But at the time he could make nothing of that. Next to Lenin and Trotsky, Pascal lacked heft.

The family grasped immediately that this was what everything had been for: the days and nights of political argument; the helpless feeling as the war closed around them; their growing distrust of Marxist abstractions as it became clear that the Party had decided, for "tactical reasons," to abstain from the battle for Algerian independence (the main "tactical reason" being a broad streak of anti-Arab racism among French workers).

In their apartment overlooking Paris, where they drank tea on the living-room floor, and watched the finely etched bulk of Notre Dame rowing each night at the same place amid a sea of walls and chimneys, they had all felt the same frustration. The veins of revolutionary energy had been stopped up while the war lapped about them with increasing savagery.

He was impressed by the friend who approached them on behalf of the F.L.N. She had always seemed so proper and fragile to him, like a character in a sentimental novel. Her hair wound severely into a bun, her old-fashioned clothing, her voice pitched with a sort of porcelain elegance made her the last person one imagined slipping through the clutches of the police on clandestine errands. Yet that is what she had been doing for months they discovered, and the discovery shamed them. Yes, they would help! They would go clandestine as quickly as possible, risking whatever the police let them risk: not their lives, perhaps, their skins were not the right color for that, but their freedom, their jobs, their mug shots and fingerprints, their credit lines.

He never learned how the arrangements were made, but they were made quickly. The apartment on the Rue du Cardinal Lemoine had an extra room which they were asked to keep at the disposition of the F.L.N. A few days later the doorbell rang. In the hall stood an olive-skinned man wearing dark glasses and a trench coat. He was in his early thirties and had about him an air of precise reserve which was so unvarying that it came, finally, to resemble a disguise. He pronounced the password, and the phrase was like a spark floating instead of leaping between them in the stairwell.

His name was Daniel, obviously a pseudonym meant to conceal his unpronounceable Arab identity; but also to soften the alienness which, he had learned by experience, Europeans were somewhat reluctant to welcome into their homes. He was the most urbane revolutionary they had ever met. And when he was installed in his room at the end of the hallway, overlooking Notre Dame and Montmartre in the distance, he became the revolutionary with the best view in Paris, although they could never tell whether he noticed this.

During the month Daniel stayed with them he rarely left the apartment, except for an hour or two in the middle of the afternoon, when the crowded streets made a police roundup less likely. For the sake of security, they agreed to stop seeing friends engaged in any militant activity that might attract the attention of police. These isolating measures created a sensation of calm and immobility which they experienced, oddly, as a form of intense activity. Their lives, quieter than ever, seemed full of furious movement. The violence still lapped about them, but now they were part of its churning. Shopping for food, going to work, buying a newspaper became acts of war, or so it seemed.

Daniel had asked them to keep alert, in case they were being followed. The boy formed the habit of walking in detours before coming home. He glanced at the reflections in storewindows, and stopped regularly to tie his shoelaces, while looking subtly around him. The anonymity of the streets seemed watchful and dangerous.

When he walked past crowds of Algerians standing along the Rue Mouffetard, or in damp alleys on the Left Bank across from Notre Dame, he felt different emotions: a profound pride of solidarity, but also fear, wariness. He imagined the knives folded in pockets. Personally he believed they had every right to respond violently to the violence inflicted on them every minute of every day by the racism of French society. Yet what would he do if the violence were turned, suddenly, against him? Would he cry out: "Stop, we are brothers"? How would he justify the hatred and terror he would feel as he tried to fight off his attackers? He imagined scenarios in which he somehow managed to explain that he too was a freedom fighter, that he was one of them. The knives would fall; they would hug each other clumsily in the dampness of the narrow street, under bulging walls which seemed to meet above their heads.

On the whole Daniel kept to himself. Sometimes he agreed to have tea with them, and then he would talk amiably enough, but almost never about the war, and never at all about his part in it. At first, one couldn't help superimposing Daniel upon the execution in the Rue Mouffetard. It was hard to imagine a connection between this urbane, sensible man and the act of murder. Yet he, or someone like him, had made the decision which pulled the trigger, and had made it for solid, arguable reasons. Maybe the victim had been a police spy. Maybe he had threatened to betray a secret. Yet the boy could not bridge the gap which separated the violence he had witnessed from the world of principles. Daniel did not seem to be the sort of person who pulled triggers. Yet, on second thought, his reserved manner had a quality of danger about it. One sensed in him a man who had trained himself not to feel a connection to individuals. When, after a few months, he had to leave the apartment very quickly, they watched him run into his room, throw some papers into a sack, and go without saying a word. In some sense, they hadn't existed for him. When he thought about it much later, it seemed clear that Daniel had been precisely the sort of man to "pull the trigger" with his own, or someone else's finger.

Neither while he stayed with them, or later, did they find out what Daniel actually did. It would have been indiscreet to ask, given the circumstances; more than that, it would have seemed like a breach of revolutionary faith even to be curious. Their job was to accept Daniel, and they did so solemnly. His presence among them was all pervading, like a note struck from a tuning fork.

Nonetheless, there were some things they could not help noticing. Daniel often came back from his afternoon outings with a fat briefcase. The desk in his room was piled with documents. The boy wasn't sure if he was relieved or disappointed when at last he connected these scanty observations, and concluded that Daniel did not go down in the afternoon to rub anyone out; that he was probably an official of some sort, with high responsibilities, which was why it had been important for him to have a stable hideout.

The war had long since entered a period of stalemate. On both sides were only victims about whom less and less could be said. The result was a sort of moral numbness which made even death, torture, and humiliation seem boring. Their clandestinity too became routine; it resembled nothing so much as a comfortable bourgeois existence. Daniel's presence had the effect of suspending the internal warfare which had been a way of life within the family. Their private struggle had been dwarfed, and then replaced, by the solemnity of the larger one.

Daniel left suddenly and, as I have already indicated, without a word. One morning the radio announced the arrest in Lyons of several people who had been caught transporting a valise for the F.L.N. Michèle mentioned this to Daniel, who leapt to his feet and ran into the hall to use the phone. He was so nervous that he had to begin his number over several times. It was the first time they had seen him display an emotion that was not circumscribed by urbane politeness or some hint of irony. His nervous hurry seemed out of place in the hallway of the apartment, where they had all walked on moral tiptoe since his arrival several months previous. Within minutes he was gone; they never saw him again. By late afternoon, several friends were in jail, and the family waited for their turn. For some reason, it never came.

At first they supposed that their policy of isolation had worked. It was not long, however, before they noticed large men in wrinkled gray suits following them around the city, and heard a peculiar series of clicks whenever they spoke on the phone.

All in all, he wouldn't have minded going to jail. It occurred to him how useful such a crisp narrative detail would be to a biographer. But this was not to happen. It had been decided in some high sphere not only to leave him be, but to spoil the drama of the moment by assigning an archetypal dumb cop to follow him around: bulky, conspicuous, probably flat-footed too, the plain clothesman who trailed him resembled a character in some terribly ordinary comic movie, not a feature in a poet's biography.

The next weeks were experienced as a form of suspended animation. The war wheeled clumsily toward negotiations. The large men in wrinkled suits continued to follow them around. The F.L.N. lay low. The jailed friends, after a desperate few days, seemed more or less settled in the comradely atmosphere at Fresnes. It was as if a fist had let go, and now, instead of falling, they were floating downward in a great lazy arc. A few days after Daniel's flight, the myth of the efficient revolutionary received a dent, when they discovered a cache of documents stuffed under his mattress and forgotten. Had the police raided the apartment, the documents would have fallen into their hands. Michèle placed the papers in a sack, forbade anyone to look at them, and hid them behind some bricks in a corner of the cellar. They are probably still there, for the F.L.N. never showed any interest.

XII

THEIR WORLD had started crumbling back to peace when history gave the family its most frightening knock. As it became increasingly clear that France was determined to pull out of the war in Algeria at almost any cost, including Arab independence, the colonial population went into a trance of despair which expanded to nightmarish proportions. Overnight, terrorism became a form of nihilistic self-expression. Flaming gasoline trucks were pitched over cliffs into the Arab quarter of Algiers. Bands of teen-age boys roamed the streets killing any Arab they found on their way, as well as any European who tried to stop them. From Paris, it seemed as if a collective madness had seized hold of the *pied noirs*, as they were called, causing them to cry "Long live death" as the last act of their collective identity.

A terrorist organization called the O.A.S., with clandestine sympathies in the army and the police, raised the possibility of a civil war in France. All of this became uncomfortably personal when bombs started going off in Paris as well. At first the victims were government officials, newspapers, and leftist politicians. It soon became apparent, however, that neighborhood accounts were being settled too. Ordinary militants were being maimed, apartments were destroyed, cars were triggered to explode. The family began to get nervous. The O.A.S., it was rumored, had access to police files, and the police knew exactly where to find them. Whenever they heard noises in the stairwell outside their door, they flinched and looked nervously at each other; or they ran to the opposite end of the apartment and waited. There wasn't much to say. There either was or wasn't a package of gray putty leaning against the door, attached to a miniaturized electric fuse. It was usually about half an hour before someone went to look. He and Michèle slept in the room nearest the door. As a result, their nights were spent on the edge of insomnia, feeling a little foolish, because their fear had a quality of hallucination, but also feeling vulnerable and unheroic.

By now the large men in wrinkled suits had begun to leave them alone. Still, their street-watch continued, except that an element of genuine fear entered into it now, a faintly freezing sensation whenever he seemed to identify a casual but persistent figure trailing behind him. Had it been the same figure yesterday, or this morning? Had it turned all the same corners he had, in the baroque pattern of evasions which he varied from day to day? He never could make up his mind.

One Sunday morning the doorbell rang and he stumbled, half-asleep, to answer it. He squinted into the early sunlight silhouet-

Automythology

ting two people he had never seen before: a dark-skinned man in a trench coat, and a girl with curly black hair, wearing a short skirt.

"Is this where Monsieur Paul lives?" the man asked.

Before Monsieur Paul could think of anything evasive or angry to say, he had mumbled yes, and was beginning to wake up.

"Claude sent me," the man said. "Are you the one who is leaving on a trip?"

They were looking at him awfully hard and long, it seemed. He hesitated for a moment, then mumbled that he didn't know what they were talking about, and shut the door. Climbing back into bed it dawned on him that it was six in the morning, and he had just had a very odd conversation with two complete strangers. These strangers had apparently not rung the wrong doorbell, since they had asked for him by name, if Monsieur Paul, sinister and gangsterish, was indeed his name. He thought about this for a while, then he woke Michèle up and they thought about it together. The only explanation he could think of drained all the strength from his body. The O.A.S. had become bored with bombs, he reasoned. They had decided to graduate to the more exacting level of personal assassination. These two people had been sent to get a good look at Monsieur Paul, so that when the time came they would shoot the right person.

The explanation made no sense at all. It was grotesque and irrational, therefore quite on a level with the nihilism of the O.A.S., which is to say, entirely possible.

How did one deal with such an extraordinary state of affairs? The fact was, he possessed no mental frame within which his assassination made the slightest sense. It all seemed terribly fictional to him. While he put his clothes on he mused, with detached panic, that it would continue to seem fictional even as the bullet entered his body, even as the pain spread from the point of shock.

He talked the situation over with Michèle, and they decided what they ought to do. Of all courses of action, he realized later, it was by far the worst, but it had the advantage of breaking right away the almost comic tension they felt. The dark-skinned man had been sent by Claude, he said. This did not seem likely, but it was worth verifying. Claude was a member of their clandestine group. Although he lived a few blocks away, they hadn't seen him since Daniel disappeared several months ago and the police had started to follow them around.

He finished getting dressed, closed the door behind him, and started down the stairs. His footsteps creaked in the chill space of the stairwell. He imagined the mysterious couple lurking behind every corner. Floor by floor he stepped, looked, and stepped: no one. The building formed one side of a courtyard connected to the street by a narrow alley about fifty yards long, bordered by ten-foot walls on both sides. He glanced at the cars parked in the courtyard; a cool, silvery light clung to them. He had taken a few steps along the alley when he heard a car motor start up behind him. He knew who it was. There was no point in running, or turning around.

An amazed clarity possessed him as he continued to walk along the tiny sidewalk toward the street which he saw before him, beyond all measurement of distance. The tires crunched in the courtyard as the car came closer and entered the alley. He could feel the eyes of the pair concentrating on his back, and found himself wondering, with dizzy curiosity, if they planned to shoot him or run him down. Both alternatives seemed oddly buoyant to him. Still he walked, as if the tap of his shoes and the swing of his arms were essential to the enormity of the moment.

He didn't think: "I am about to die." He felt a glassy limpid tension gathering within him. In seconds the car had filled the alley with its roar, had approached, and had gone by him. A man and woman were in front, two children in back. It was a neighbor making an early start for Sunday in the country. He kept walking. He was halfway to the street now. The mottled gray walls loomed over him in the silvery quiet of early morning. Nothing had happened; nothing at all. And yet he had died, by mistake; he had died, and then issued forth again, expelled into the luminous tranquility of the alley. Once again he was merely alive. Once again he had time to place his feet one before the other.

Automythology

> Connecting between Paris + earlier

He felt as if an immense ball had suddenly unravelled, and lay in disorder about him.

The explanation for his "death" was really quite wonderful, as he found out later that day from Claude, who had also been visited by the mysterious couple. The F.L.N., it appears, had mislaid a valise. Like Daniel's papers under the mattress, it had been left behind in a hurry; no one seemed to remember where. So they had decided to send an agent to visit every possible contact, and offer the password. Sooner or later, they reasoned, someone would recognize it and hand over the valise, filled, he imagined, with a few hundred thousand dollars, or a packet of crucial documents, or maybe simply some dirty laundry. All in all, it had been a good day to die, and a better one to be reborn.

He felt as if a glass nail had been hammered into his life. That it had been a mistake was incidental, and only slightly comic. The nail remained, like a fulgurite in the desert, drilled by lightning to a great depth, resembling vertebrae made of hard glass and fused together.

One day in a *oued* near Taghit in the northern Sahara, looking for neolithic implements which lay about for the finding, he would come upon a long fragment of fulgurite in the sand where, several millennia ago, the sky had stuck a finger. It was glossy and delicate, compared to the clumsy fragments of flint which some neolithic savage had hammered with small success. The fulgurite was almost decadently beautiful. It reminded him absurdly of the "glass nail" in Paris that day when he had "died." The "nail" had served as a marker along the track stringing one flatland to another in his life, one yellow blindness of sand to another.

XIII

THE WAR soon was over, and the peace had begun. In some other country a President was assassinated, launching a nightmare into which the boy would enter when his time came, toward which, even now, he progressed. For these years of the elegy, like Hieronymus Bosch's globe, had come to seem ever so faintly hollow, as if all this time he had been almost-living and almost-doing and was beginning to find it out. He was, in a sense, already going home, and had been from the day he came to Paris.

Michèle it seems had always sensed this. She had tried to imagine a life for herself in New York and couldn't, and had become quietly bitter. She let him know that she knew, and his response had been anger, resentment, for he was not ready to accept that his impersonation would come to an end.

He never really thought about going back to America. From the first, his life in Paris had seemed complete. The tender if delayed jostling of his emotional life, the mature shadow which had begun to sculpt the angles of his face, had fed heartily on the elusive food of his foreignness. When he dropped pebbles into the well, the echo from the deep was French. His sentimental tendencies were mirrored in the lyrics of Leo Ferré and George Brassens. When he sang in the shower his voice twinkled gravely with the tones of Yves Montand. He welcomed the raw slopes of the Alps and the poplar-lined streams of Île de France like a homecoming. He had memorized the link between *boudin* sausages and being broke. His fingertips understood the feel of Camembert cheese: which were ripe and pungent, which rubbery and bland. He drove his car—an affair of bolts and tin known as a 2CV, a later avatar of which would carry him through the desert—in French. He loved and hated Michèle in French.

Yet there were signs, still mysterious to him, that an Archimedean point existed which would lever him out of this world; that its wealth one day would appear to him, psychotically, through the small end of a telescope as he hung between equally distant planets: one finely inscribed with the contours of a decade, France; the other awash in elemental tides, unknown, yet achingly familiar, New York, to which a ghostly thread had linked him all his life.

As yet these signs were cryptic. He was uneasy when people assumed that he had become a French citizen. To tell the truth,

Automythology

he had never even considered it. He was surprised at how reasonable the assumption was, given the uncanny perfection of his French, but also at how repulsive it seemed to him.

It was important to keep his balance on the high wire. But it was also important that a net be there to catch him if he should fall; a secret sharer to whom he could turn when his life became overly serious, whispering: this is not real; this is happening too far from home to be real.

Another sign was harder to grasp because it was conveyed by a feeling so diffuse he scarcely had a name for it. He was lonely. ✓ This is not how he expressed it to himself. He imagined over and over again a friend he could go drinking with; someone he would not have to talk to about anything in particular, yet the ordinary complicity of talk, in a café over a cognac, would be resonant and warm. He did not have such a friend. He had acquaintances who were available for lofty conversations. He had ideological partners—he called them fellow-travelers—with whom he could exchange dinner invitations, borrow books, share climates of opinion. He liked these people. He liked their world, their way of thinking, and their style. But when he applied his test to them, they failed. They wouldn't do over a bottle, even less staggering down a street. They wouldn't do when nothing was doing but friendship, its model surviving from ancient days: a college roommate he had been so close to that most people never got their names straight; childhood friends in Brooklyn with whom he had cut school and horsed around with tender vulgarity. His models of friendship were paleolithic. They dated from the street-wise days of his boyhood in Brighton Beach, when intelligence and spiritual finesse had been considered, at best, to be impolite.

There was a contradiction here which undermined his self-esteem, for the better he became at the adult skills he valued, the further he drifted from the inarticulate fumbling love which had bathed his childhood: the wise love of his grandmother, whose broken English was the exact opposite of his perfect French; the profound, anxious love of his mother, at home mainly in clichés; the remote love of his father, which rarely got further than his heart, leaving his face and his words in a silence which had wounded the boy, but had provided him with a lifelong envy of reticence, which he associated with dignity, and valued passionately in the woman he eventually would marry, after returning to America and the anecdotal world.

It would not be entirely wrong to say that he came back to New York after ten years to see if he could find a friend. Meanwhile, he didn't say he was lonely, because he was not in the habit of naming qualities of experience. He said: "I don't have a friend," thinking, "Maybe I'll find one."

A few years later, still in Paris, talking with an older poet who would help him take his first timid steps toward home, his experience of himself would be changed by a few casually said words: "Loneliness is when you want something very much, and can't find it," he had proposed. "You feel a lack, and look around, and no one's there."

The older poet's wife stood up angrily, and shouted at him: "Wait a minute. Loneliness has nothing to do with exterior things. Has it ever occurred to you that the lack you're talking about is in you, not in the world; that there's a difference between solitude, and loneliness?"

No, it had never occurred to him, but it did now, suddenly, luminously. A feeling of luxury possessed him; a profound gladness which smiled at him from the rug and the dented couch, from the peeled ceiling and the gently inadequate lights which made the room seem almost gloomy. Quite as suddenly as he had discovered Communism, years before, on a headachy hillside in Provence, he discovered now its antidote: the inner life; the notion—entirely startling to him—that something of interest might be taking place in the claustral reaches of his mind; that certain experiences which he had previously brushed aside as forms of inner static might actually be the sources of everything: like a bottomless, crumbling soil, a home for roots.

As the Algerian War receded, and the pulse of revolution no longer throbbed under his fingertips, he found himself suspended in a windless space. He flailed his arms and legs like an astronaut in quest of gravity, slowly, insectlike. The space was his loneliness, which his innocence still perceived as a quality pertaining to physics and geology, not to emotions.

His loneliness took another form too. He became frightened by his enormous facility with French. At times he felt that he was possessed by a monster who spoke when he spoke, smiled when he smiled. He reached out a hand, it was the monster's hand. He lay back and watched the alien creature. He felt at times that he was its victim.

Once, in February, he went to London for a week. He visited the necessary sights, but mostly he walked, went into shops, eavesdropped on conversations. He was avid to hear English spoken; even this odd version of English which, in his absurd mythology, he connected directly to the speech of butlers on the radio shows he had listened to when he was a child. He almost wept with pleasure, hearing people ask for groceries in English, gossip in English, yell at drivers in English.

For years these circuits had been buried and planted over with a garden of vicious elegance. He didn't like London, but he liked to listen to it talk. It poked open the shutters in a long-closed house in his mind. A mental breeze lifted the dust from the furniture. Ghostly shapes opened and closed doors, flushed toilets, and wrote letters. Gradually the house became less ghostly, as if another life were developing itself in the shut-down space; as if he were beginning to have in place of one fictional existence, perfect in its way, two antagonist existences, one of which had prior claim, for it was the reality he had locked away years before. And it spoke English.

XIV

ALREADY he had let go many of the connections anchoring him in his Parisian life. He had become self-preoccupied but didn't know it. Instead he felt two things: a sort of moral low ceiling, as if the sky had been removed, and something dusty put in its place; a transparency, born of inattention, which made it hard to concentrate on other people, so that most of the time they appeared slightly out of focus to him.

He had begun, imperceptibly, to live by rote. This did not disturb him. On the contrary, the mood it fostered had a perverse, almost seductive, resemblance to peace of mind.

The image which expressed this peace of mind lay at his feet one whole summer: an oval sapphire set in a perfect cone of dusky green. It was the Lake of Nemi, located in a hilly region south of Rome, where he and Michèle spent several months around this time. The village of Nemi was strung along a cliff overlooking the lake. Each morning, he faced the lake from the narrow balcony where he wrote. The writing served as a lens magnifying his awareness of the lake. Its jewellike immobility fascinated him. When there was a wind, a rug of silvery pleats intensified the immobility, which held him in a state of ice-cold musing, morning after morning.

In the afternoon, they would row into the middle of the lake and go swimming, or lie in the boat and read. One afternoon, the woods at one end of the lake caught fire. Columns of flame curled up the cliff and licked at the houses in the village, without quite reaching them. From the middle of the lake, the flames resembled great orange flags. The pageantry was barbarous and splendid. For these woods formed the sacred grove Frazer had written about on the first pages of the *Golden Bough*, where the aspiring priest of Diana Nemorensis prowled, knife in hand, in search of his victim, knowing that he too would be the victim on another day; that he would be torn from life by the very weapon he now wielded against another. From a distance it seemed that the rocks of the cliff and the waters of the lake were burning.

Shortly after this, the boy's self-preoccupation caught fire too. It started with a small, dull blaze on the shores of his peace of mind. When it ended, he was living alone in a damp room on the Rue Maître Albert, trapped in the amber of insomnia, while he

Automythology

tried to decide whether he had been crushed or exalted by the cataclysm which had blown Michèle out of his life.

For years their existence together had been ruffled and angry. They had turned struggle into a private heroism which took the place of love, because it let loose, painfully, desperately, intensities of emotion which would otherwise have been lost to them. The modulations of struggle—physical, mental, sexual, spiritual—became an instrument they played on, secretly proud of their fierceness, thinking nobody else could take what they took from each other and live. They had driven themselves into each other's lives like great steel spikes. They were bound to each other along molten seams, like girders run together after a heat bomb.

All of this made for a kind of purity which fascinated people. They were angry pilgrims: he intense and translucid, his arms loaded with all the volumes of Hegel, all the volumes of Freud, and a thin notebook of blank pages; she blond and Wagnerian, creating on thick sculpted canvasses a portrait of desperateness in cruel slashes of color, as if the id were peeking through a window.

Around this time the question of marriage came up. It seemed like a strange subject to him. Not that he was against it. He had simply assumed that their relationship was complete; was, in fact, so crammed full that he couldn't imagine where they would fit a marriage. Michèle remarked the abstractness of his interest and resented it bitterly. The fire started when she decided to take her revenge by having an affair. She picked the ugliest man she knew and announced it that very evening. There was a nuance of curiosity in her voice, as if she wondered whether he would notice what she had said.

He noticed. Every sexual hair on his body stood on end. The sacred grove crackled and roared, each tongue of flame a ghost of the goddess licking his raw nerves. He got out of bed and stood with his nose to the window. The bulk of Notre Dame rowed in the darkness. Off to the right, the clock tower of the Gare de Lyons loomed discreetly, resembling a Gothic penis.

Now that he thought about it, the night was full of penises, including one which had screwed his woman a few hours ago. The window pane pressed against his skin. He felt a pressure building within him, a feeling of potency mingled with despair. The despair was enormous. It streamed in a steady outward gust, blinding him. Yet, under its poisonous velvet, he was surprised to feel an erection pressing its whole length against the glass. For the first time in his life he felt a desperate sexuality pouring through the world. Paris, speckled and vast beyond the window, was a sexual clam. His feeling of pain and betrayal; the fat pimpled face which had screwed his woman; the woman frightened and awake in bed, waiting, staring at his silhouette: everything siphoned in one velvet swirl into his penis which stood on end and leaned against the window.

They wept and made love until the first glaze of dawn embarrassed their isolation. But maybe it wasn't love they made. Maybe they were trying to touch something alive in the suddenly mineral night which had turned cold and quiet, and had been quiet all along, but they had not listened. After each climax they fell apart and lay in the chill which drove itself against them and drove them against each other.

By morning they had guessed that they couldn't comfort each other any more. They dressed and began the day, and decided it was an ordinary day. He would not pack his bags. She felt sick when she remembered the pimpled face of the man whose penis she had invited inside her; she would never see him again. They did not announce their decisions to each other. Such things went without saying. Yet their relationship was over. It was over because he had looked inside the walls of her existence, had wept and trembled with emotion, but he hadn't come any closer; he had ignored the anxious surrender of this woman who wanted only a touch, a signal of something soft and hopeful.

So in the morning they agreed that nothing had happened. She scraped her painting knife and squinted at the anger of her latest canvas, not finding in it the tenderness she never knew how to include in her artistic language. From across the room he watched her and went on getting dressed. He felt as if a claw had been hooked into his body all night. The bed clothes were twisted into ropes, making him think for the first time in months of something that had occurred the year before. They had come home one day and found their bed strewn across the room: the mattress leaning against a wall; the sheets, blankets, and pillows trampled on the floor. They hadn't spoken about the incident to anyone, but they were sure they knew who did it. It had been Michèle's youngest brother. They could almost smell the tight-lipped anger lingering in the room. They could almost see the controlled frenzy of the handsome adolescent methodically tearing the room apart, with an expression of amused irony on his face and an abyss of violated "horde" pride in his arms and stomach. For the first time he thought he understood the sexual ache which gave rise to such explosions. The stuffy mystery of the bed lay beside him, innocuous in daylight. The intimacy of the used sheets seemed timid and ugly to him now. He had been exposed in the sweaty shadow of this bed. He had been driven a little crazy by its smells and its oddly comfortless softness. It occurred to him how solitary the night had been.

He glanced at Michèle. In a musing sort of way, he sensed that she had forced this climax not simply to punish him, but to bring about a change in their lives, almost any change would do.

This became clear to him only later, after they had broken up, and, happy in his aloneness, he began, for the first time, to understand the complexity of her character, her desperate generosity; began to understand them because he was not flailing about in them, girding to survive the bewildering alternation of love and disdain which she flung about her so magnificently.

The idea of punishment was especially closed to him at the time. He hadn't done anything to be punished for. As far as he could tell, everything was perfectly all right, and he was all right. They were all all right. She had done something terrible and strange, like an act of God, and now they were picking up the pieces. It didn't seem to him that he even blamed her. The important thing was to get back to normal as quickly as possible.

Three Journeys

The earthquake subsided more slowly than he would have liked, and he himself, unaccountably, was the reason. He found himself running around frantically trying to have an affair. He offered to simplify several previously ambivalent relationships with women-friends, but the women, sensing his one-track mind, preferred the mystery of ambivalence to the conclusiveness of a love match. He hung around the Select Café in Montparnasse, and finally made headway with a plump German girl who was living at a nearby hotel while she looked for a job as an *au pair* girl. She was frightened by his grim insistence, but she was lonelier than she was frightened, and they went back to her room. He was so methodical and pleasureless in his attempt at seduction that the girl finally pleaded with him to leave. To his own surprise he gave up rather easily. He left, and a while after that gave up the idea of being unfaithful.

It took some time for the breakup to come, mainly because he was not very helpful. He insisted on being content with their life as it was. They didn't speak of marriage any more. In fact, they skirted all the darker patches in the landscape, becoming each day less visible to each other, less volatile. He had reverted to his earliest landscape, and she, a little confused, let it happen. She trailed him into his desert and kept her mouth shut, because she sensed his regression and was fascinated by it. He had not simply "turned off." He was wandering around in the "ocean made solid." The absence in his face was not just indifferent, it was childish, almost moving. As if he were about to cry. She tiptoed around him, not knowing what would happen next, but beginning to guess that whatever it was, she would have to make it happen. It is possible that, for the first time, she began to love him during this period, as one loves something which is beginning to end.

In truth, they had some of their happiest times during these months. Experimenting with his new Americanness, he had made several friends. There was David, a spectral fellow from Philadelphia who wanted to be a writer, and felt that he owed it to his muse to experience every variety of low life Paris could offer

Automythology

him. He drifted from a nebulous relationship with his one-legged Russian landlady to the indecipherable cool of jazz bars and all night cafés, gazing with fierce gentleness at the creatures inhabiting what he looked upon as his private underworld.

There was also a black Jamaican writer named Lindsey, who spoke with a sculpted lilt, as if he were on the verge of singing. He wrote stories filled with lullabies and outbursts of crazy violence. His stories melted into his voice, and his life was modeled on a myth. In his face one saw grass huts, dope, and lovemaking in the yellow cream of Jamaican moonlight. He walked among the streets of Paris like a folktale written by Rimbaud: full of forced laughter, innocent, but expressing his innocence as a perpetual flirtation with suicide. One was never sure if Lindsey would survive the day, or if he would live the myth to its conclusion, stumbling under a truck on a drunken afternoon, or getting mugged to death under the bridge where he insisted on sleeping.

David and Lindsey were each other's white and black shadow in the crowded bookshop near Notre Dame where he first met them. The two experienced each other in a purely mythological way. To David, Lindsey represented youth and life, and a gaiety uncontrolled enough to be murderous; to Lindsey, David represented the most intense passion of dignity, a decanted presence which one was tempted to call angelic, except for the restrained rage peering from his eyes on occasion.

Somehow he was accepted by them as a third. They spent a great deal of their time reading manuscripts to each other and eating in scruffy Arab restaurants. But it wasn't literature that bound them together; it was talk along the dark muscular Seine; talk in the musty paradise of the bookshop's upstairs parlor, lined with grimy bookspines and gnarled quilts draped over an ancient, crawling mattress; talk on the floor of the apartment on the Rue du Cardinal Lemoine.

For something had happened which he could not explain: Michèle accepted these friends. Instead of the suspicious irony he had been prepared for, there was laughter and hospitality. A sort of youth possessed her which he had never noticed before. After a while, the threesome became, in fact, a foursome. Their friendship was intense and lovely, and—strangest of all—it passed the "drink test." Although the test this time was applied in terms of pot, its bilious green flakes becoming the heraldic sign of their four-sided happiness.

They smoked pot for the same reason that shepherds play the flute in Watteau's pastoral fantasies: to weave a social magic, casting a silken net over the nameless flowers and perfect soul blue of the sky. Their smoking was sensual and innocent. A mirage in an empty place? Perhaps, but a mirage you could taste and lie down in.

They spent afternoons at the apartment, listening to Charlie Mingus' *Pithecanthropus Erectus,* Miles Davis' *Friday Night at the Black Hawk,* or Thelonius Monk shovelling clumps of chords from the scratchy record Lindsey carried around with him as some people carry around a book. They ate jam, drank tea, and did funky word-solos as the smoke curled in their lungs, raised its gem-studded head into their eyes, making them sway with underwater clarity, their gills expanding, their hands falling over each other and over each other's hands.

After each smoke, he wrote poems all day. Or rather, he broke off lengths of poem from the recitatif which poured, hour after hour, into his pen without seeming to transit in his mind. Standing on subway platforms, or lecturing in front of a class; having breakfast, or making love; he was possessed by his song and wrote it down, or else, with awesome wastefulness let it drain away and be lost. The poems were vapid and unfocused. They were not poems at all, but fossils left over from the tropical age of smoke.

If there was ever a time in his life he could point to and say, "I was happy," it was that time; their four-sided mandala fused into an aquatic unity of faces, arms, tea on the floor and talk in the smoke-slowness of the streets, in the amber eternity of those late afternoons, sharing the wrinkled stem which only Lindsey knew how to roll; the boy presiding bizarrely with outstretched Biblical arms, and a talent never-before-or-since-known enabling

Automythology

pot - mind
journeys back
in time -

him to speak poem-solos of Yeats or Walt Whitman in perfect synchrony with the bucking saxophone of Charlie Parker.

The English language, too, began to put out frail shoots. The playfulness of pot caressed his locked-away images of childhood and adolescence, kissed his private parts of intimate remembering. It flushed their ghostliness with the youth of tastes, smells, touches which his invisibility had shelved and forgotten, but which had adhered to the words of his mother language coming back to him, fleshed and playful.

When, a long time afterward, he tried to reconstruct a time for this intrusion of happiness, he would be surprised to discover that it could not have lasted more than a few weeks. After that, summer had come. David had gone back to America. Lindsey had begun his gradual drift into hatred, recasting his Rimbaldian folklore into a less lovely racial folklore which culminated one night at the Café de Seine. Lindsey had come to him through the crowd at the bar and grabbed the back of his neck with one hand, shouting and weeping that he loved him more than anyone in the world, more even than his father, but that he wanted to kill him, and maybe would kill him some day, because he was a honky. That had been a low point: that packed, narrow bar with its pinball machines and its drifters en route between Amsterdam and Tangier, with its bizarre warmth which he pulled around him like a found coat in the miserable February of his loneliness, when he marked off twenty days on the calendar during which he was willing to swear that he never slept.

Those wonderful weeks of smoke and loving, only a year before, had marked the end of a sentence, full of parentheses and subordinate clauses, which it had taken five years to arrange syntactically so that its last fall would have the effect of a New Orleans jazz band playing at a funeral.

Those weeks had caused him to teeter drunkenly in his lean self. They had instructed him in a new possibility which would unbalance his lifelong effort to stand systematically still in his bones: the possibility that he might give himself pleasure, that his daily acts might be ends not means; that his imagination might steam and bubble for the hell of it, and that he, at moments, might exist for the hell of it. After that encapsulated lesson, more like a crash course, they had begun their speechless drift to the bottom of the page. It was all so amazing in its ordinariness.

*   *   *

The elegy doesn't end so much as it runs out of room. Reality begins to crowd it: a marriage beginning and ending; another man's shrewd silence, and his patience; the boy sprawled against his will in the anecdotal world, forced for the first time to match his interior aloneness with actual solitude.

The other man was François, also a painter. He had apparently been working on one painting for a year and a half, and the idea of so much integrity stunned Michèle. The painting had been so browbeaten by the artist's integrity that it had come to resemble a slab of raw meat. In fact, through some sort of cosmic ambush, it looked awfully like some of Michèle's own paintings.

The geometry of their relationship was muted, triangular. They seemed to be trying to accomplish something without knowing what it was. The boy simply took a deep breath and one day let it out when Michèle said to him: "François is in love with me, you know."

There was something unpleasant about the calm which flickered within him during the days that followed. Why wasn't he wandering around in the streets talking to himself? Why wasn't his hair a mess and his clothing unwashed? He shifted his weight from foot to foot, and brushed off the dust, while the earth shuddered, and he concentrated on the small triumph of staying on his feet.

She was leaving it up to him, she said. She would send François away. All he had to do was envelop her with mature love, recognize her fears without seeming to, stop wanting secretly to go back to America, be tender, buy an apartment, admire her painting, heal her anxiety.

Automythology

He felt paralyzed. Yes, he would gladly do these things. But, tell me, there must be a method, some lessons.

No lessons? A deep inner change? He became more paralyzed than before in order to suppress a shout which was beginning to percolate from deep down in his being: Leave me alone. Go away. I don't love you.

The elegy is finished. There is only guilt, insomnia. Broken pieces and secret pleasure.

There is the absurd last summer when she went away to the country for a month to think it through, and he, awaiting her verdict, guessed that he had won, because waiting was his favorite activity.

For a month he gazed at the sunset from the wide windows of the apartment and took long walks in the early evening. He ended up at the bookstore across from Notre Dame, where Danish girls unfolded their lank, lovely bodies reaching for books on the high shelves. He sat in a punctured armchair, eyeing the scene with a combination of detachment and sexual curiosity. George, the owner of the shop, introduced him as an old Paris hand, a fine flower of evil. He smiled at George's manic flourishes, and the girls giggled. He was full of advice about *kif* dens, discothèques, and Arab restaurants. All of this came naturally to him, and yet it was entirely new. He was masterful, in control.

The month passed, and he played, for even here he could not commit himself. His passiveness was crucial to his peace of mind during this eternal July. Unreality was his consolation. Dreams fulfill wishes which the protagonist approves of only in that form: as the scenario for dreams. This was his situation. He dreamed the life he was living and therefore didn't live it at all. He waited.

One sees him walking, birdlike and solemn, past the battered cafés of the Rue Galande, or mingling his ghostliness with the bearded figures exchanging joints on the shiny cobbles of the Rue de la Huchette. One wonders if, during this period, he was wholly visible to others, or if he signalled his nomadic presence by some optical disturbance, some refraction of the sun's rays which for a moment seem not to know whether or not they ought to cast a shadow.

What happened next was more real than anything he had ever experienced. A day driving south at the end of July. The long, empty sunlight of ever narrower roads, ever drier landscapes. Lyons, Arles, Nîmes. The strident buzz of cicadas sawing their way out of all the jails in the world. The meager scrub land of the Garrigue: abandoned houses, scrub oak and acacia, a hot syrupy wind scented with thyme. The partly ruined village of Notre Dame de Londres, a mile beyond the end of the last, narrowest road.

Michèle was walking toward him over the ruts and the animal droppings, with her gorgeous blond hair and her spendthrift body. She smiled, and he knew how Jonah felt. That was the way it always was after a separation. A swallow, and he would be inside her, happy in his way. There was something mysterious in her smile. Before he could think what it was, the village began to poke its head out of doors, starting with dogs and children. Then came the elders who sat all day by the fire in windowless rooms, even in mid-summer. He was introduced to a venerable white-haired man named Raphael, who completed the classical reference by pointing with outstretched arm at the setting sun, and saying: "Phoebus."

Michèle helped unload the car. She lived up a flight of stairs in a pleasingly bare, whitewashed room. Her painting affairs were in a stone barn outside the village. They hugged in the almost chilly dusk, and he waited for his mind to fill with her presence; he waited for her sensuality to unfold the ritual they had perfected during six years. He caught himself waiting and wondered fleetingly why. They made love, and, when they were dressed, had some supper. Later that evening, they walked through the sheep-cropped fields near the village which had been turned into negatives by the summer moonlight.

The next day they took a drive through the Garrigue. The gullies slashed into hillsides by flash floods had a beginning-of-the-world look. The boy was excited by the harshness of the land. While they talked, he imagined himself tracking through it for hundreds of miles. He imagined waterholes with steep, rocky sides, almost impossible to find. He imagined that he knew which roots to eat. He imagined himself poised in concealment beside a rabbit warren, ready to seize the small brownish creature with his bare hands.

Michèle was talking. François had come to see her, she said. He had been emotional, out of control. It had been very hard. But she had persuaded him to leave right away, not even to spend a night.

She didn't announce her decision in so many words. She held onto his arm and talked, and he nodded. They were going to live differently now. They would move out of the apartment on the Rue du Cardinal Lemoine, and find a place of their own, even if they had to borrow the money. They would get extra jobs, she would work too. They weren't students anymore. They didn't have to improvise their lives. It wasn't enough to be buddies, they had to be adults, man and woman. Her voice was warm. She stroked his leg quietly.

So it's settled, the boy thought, blending mimetically into the landscape which flowed past the car on both sides. The rasping expanse of rock and dried bushes stilled his mind, as if a brown, pug-nosed creature had sniffed the hot breeze and frozen, become a nob of nothing in the scrubland which blinked light and dark at the sky as it had for ever so many thousand years. For a sublime instant, he knew what it was to be hunter and hunted.

He smiled. "Yes, of course," he said. "We'll live differently now."

Something broke in him. He would never forget that moment. It simply broke. A pain filled him from head to toe, and he thought he was going to die. Then he realized that the pain was not physical. It was a moral pain which started in his stomach, and sent filaments into every part of his body. He sat, and drove, and smiled. He let Michèle's tenderness move him about, find words for him to say. He was a marionette of her need.

The need was so strong. He had never understood that before. She had opened entirely to him and was scared by it. And all the while the shout was rushing from its cave, louder, still louder, a deadly subway. He didn't want to hear it. I don't love you!

The next day he told her about his anxiety. It was an ailment, he said, a loose wire. It will get better.

Her face became hard and she began to cry. He cried too. It will get better, I know it will. Besides, what is it?

She knew what it was. Now they were each other's marionette.

It was his fault, all of it. That was the part he couldn't bear, that was what crumpled him into soot. It would never make sense, and it would prevent anything else from making sense.

They got married three weeks later in Paris, along with two dozen other couples in the Mairie of the Fifth Arrondissement. They celebrated the marriage with some friends in a Tunisian restaurant on the Rue des Boulangers. After the meal, the owner grabbed everyone by the hair, and poured *raki* down their throats.

A few months later, he was found one morning with his throat slit, maybe a customer who didn't like *raki*. Everything else was ending around that time too.

It was a misty autumn. François crept under the window, oozed out of the telephone. The boy, in his cell, turned stones into skulls, trying not to hear the shout which was travelling toward him. He accomplished this by keeping out of Michèle's way as much as possible. Michèle and François helped him with the shout. They kept out of his way, and in each other's.

The Église Saint Severin, one evening in early December. Gray colonnades disappear above the spotlights. Bach groans. Beside him, Michèle is crying silently. "What are we doing?" she whispers. He didn't know what they were doing. Suddenly something very cold crept close to him. Nothing in his life had ever been irreversible before. But now, on the hard church bench, the clang of an ending filled his mind. This person beside him with whom he had lived for six years was no longer his completed part. That night they would occupy the same bed. But he was cold with the wind of outside.

A week later he moved out; pleading, scared his pleas would be heard; relieved, paralysed with the fright of change; numb, exquisitely aware. Full of ending and beginning. High and dry in his aloneness, flooded by what was merely happening.

A wooden heron hangs over a pet shop on the quai of the Seine across from Notre Dame. It has a neck like a snake, and a beak like agony. In the mists of Christmas, it stands out above the car motors and the shoppers. It has become his nightmare. The paint streaked and peeled, the neck a frozen undulation, the beak permanently cruel about fifteen feet above the sidewalk, pointing into nothing. He walks under it on nights when he can't sleep. He can rarely sleep. He's not sorry it has ended. He is merely terrified. He is surprised that he is not sorry. Yet, under the surprise, the terror, is excitement beyond words, a feeling of wonder.

The bird impales him with its snake.

PART THREE

The Bright Yellow Circus

LATELY I have begun to think about the wisdom of fairy tales, in particular the story of the poor farmer who is granted three wishes and ends up with a string of sausages dangling from his nose. The story is a reminder that it isn't enough to obtain the power of wishing, you've got to know what wishes to make, and when to make them, for wishing is a form of destiny. Making our wish, we make ourselves. We exist in the time between the wish and its fulfillment. On the whole, we are creatures of the interval. Caught between a past of wishing and a future of dreamed fulfillments, we project a sort of inner extension or continuity which we come to think of as our personal sign, our very selves. Our intimate awareness results from the habit of wishing; it is the fossil of a wish.

I began to think about wishing because, for some time, I had felt something dangling from my nose: clumsy and a little terrifying, hard to explain to friends and utterly mystifying to myself. It felt as if I had been chosen by a demon whose bizarre humor had turned my face into a nightmare, and I remembered the philosopher who said that after the age of thirty a man is responsible for his face. The idea is absurd, I said. How can I be responsible for these damned sausages? Yet I had suspected for a long time that it was true; that some recipe of wishes dating from the earliest mixture of my being had prepared this awful dish. As far as I could tell, I had intended something far different, something tasty and succulent. I had intended to be completely happy, and had set about my work with confidence. But some-

"The Bright Yellow Circus" originally appeared in *American Review* 23.

thing had gone wrong, for the recipe had produced this awful appendage. I suspected, to my horror, that I was myself the demon, and had no idea how my intentions had become so twisted as to make me permanently miserable instead of happy.

Pretty soon I couldn't think of anything else. I heard a whimper reverberating in my ears, which I gradually recognized as my own voice. What was I looking for? Was it the demon, or was it myself?

I began to understand that weird power of mind which is the subject of the fairy tale: the power to make wishes come true. Gradually our lives become filled with the furniture of our wishes, and it is too bad if we have gotten it wrong, for we may end up with sausages dangling and self-preoccupation repeating itself like a mad tape recorder.

My wishing began about fifteen years ago, as well as I can remember. A kind of alchemy had passed my elementary substance through tubes and chafing dishes, until a spectral self collected in the mental cup of a twenty-five-year-old boy living in Paris. The boy had done away with his past. He spoke only a foreign language, which had become a circumlocution for the name of God, childhood, or self. His life was a euphemism. He preferred Hegel to sex, ideology to emotions, the mathematical ballet of Mallarmé to the compassion of Walt Whitman. He had talking partners, but no friends. He was lonely, but hadn't yet learned the meaning of the word. He enjoyed the comfort of ghostliness, which enabled him to project a complete wardrobe of social relationships, including a marriage, without having to wear any of them. It was in such a space that Galileo's theoretical feather came to earth with a velocity equal to a stone's, proving that gravity is relentless and democratic.

That is how my wishes came, as if by appointment. I wished I had some friends, I wished I could write better, I wished I could find some honorable way of going back to America.

I began to remember certain moments in my childhood, like a cluster of bright balloons floating over a beach. The beach was silvery and empty, the white pleats of waves unfolding toward my feet.

Inside one balloon was my grandparents' farm in upstate New York, with its horse and buggy, its birch forest, and its sentinel pine tree on top of the hill, like a barbed green mooring for the sky. Here I first learned to pray: "May I not meet any snakes today, may I not be stung by a bee, a wasp, or a hornet; may I not be scared by a spider." The natural world represented extraordinary peace, and extraordinary terror, and the ambivalence did not disturb me.

Another balloon contained a decayed wooden porch in an alley near the ocean, a cushion to sit on, and a large number of books selected according to several principles: there had to be much natural description, empty spaces, characters preferring solitude to people, and a busy sprinkling of adventure. The perfect book, beside which all others succeeded or failed, was *Robinson Crusoe*. The experience of loss foreboded by a story's end was so distressing I decided never to read a book less than 300 pages long. To this day, the imagination still has for me a smell of dampness and rot-softened wood; an alley roofed by a strip of sunlight; neighbors' voices shouting through windows; a basement full of coal and, at the far end, the oblong opening to the street through which bus rumbles and the noise of strangers drift mysteriously.

Another balloon contained a dream I had at the beginning of adolescence, the only boyhood dream I still remember. I am lying in bed, in a large bare room with windows opening onto a courtyard. The courtyard lets me know that the building is my elementary school. Women come into the room with graceful ballet movements. They are naked. They bound onto the bed and out of the window, one after another. I know that they are bounding to their deaths. I understand that this is a nightmare, but the feeling of terror is muted. I am not sure I feel it.

Around the time the wishing began, a creature made its appearance in the dark of my character. It shadowed me from within, loving when I loved, speaking when I spoke. Every spoonful of my existence went, somehow, into its mouth. This creature burgled me relentlessly. Because of it, everything went wrong: food did not feed me, but it; success did not please me, but it. The creature reclined in the sultriness of my inner existence, while I shivered and became thin. Oddly enough, I was aware that I had made a bargain with this creature, who was none other than the demon of wishing. I would get my wishes, but the satisfaction would be his. As a result, every wish would make me hungrier. Every day I would run faster, but with increasing difficulty, as my inner accomplice grew sluggish and fat. I could see the day when I would fall exhaustedly on the ground, and say: "That's it, now it's your turn to carry me." Meanwhile, running and hungering, I squandered my pleasures on the demon.

I think of Walt Whitman sitting under an apple tree on Long Island at the age of thirty-six, drunk with the odor of crushed grass. I think of William Blake conversing with angels, and Jakob Boehme cobbling God's shoes. I think of my experience of the Sahara Desert last spring, and of another more surprising experience only a few months ago, when I received a phone call from a friend I hadn't seen in years. At the time, she had been about to set up a psychiatric practice somewhere in southern France. After some hard living, that was what she had decided to do with her life, and I remember feeling relieved to hear it. It was good for people to pull themselves together, I thought. A career was like a net over the abyss. I felt that for myself and needed, I suppose, to be reassured by the same reasoning in others. The contrary choice generally frightened me.

"I've been in India for the past three and a half years, living in an ashram," she announced, "and now I'm in New York for a while with my guru. Why don't we get together?"

It was unsettling news, to say the least. Apparently my friend's net had become unstuck since I last saw her. I was embarrassed to admit I didn't know what an ashram was. My friend had gone to India to "shop around for a guru," she explained, and after some looking had found one. I didn't exactly know what a guru was either. The image that came to my mind confused the Aga Khan sitting cross-legged on a pile of gold with a plump Hindu adolescent grinning blissfully from posters all over New York. Gurus had something to do with the wisdom of the East. They were some sort of wise men you went to when you had a question to ask. I concluded I would rather ask my friends, or read a book.

We had coffee together the next day and she talked about her life in India. She had always been a tough-minded lady, and that hadn't changed. If anything, she seemed tougher now, almost ominously solid. Whatever she'd been up to, it had somehow accentuated her personality. She spoke slowly, pausing for a long time between words. I had the impression she'd never talked about these things before. That struck me more than anything she said. What sort of experience, I wondered, could a strong-minded, intelligent woman have been engrossed in for three and a half years, apparently without having tried to explain it, even to herself? A few other things stuck in my mind. Her guru, Swami Muktananda, wasn't simply a guru, he was a Sadguru, the highest level of guru; so high, he had nothing in common with the pleasant-faced Indian gentlemen in white clothing one met presiding over yoga centers throughout America. A Sadguru was not simply a teacher but a "perfectly realized human being." I heard that expression in quotation marks, because it didn't seem to me it could be used in any other way. "Once upon a time there was a perfectly realized human being. . . ." What do we mean when we use such an expression, I would ask my students? What in our psyches responds to mythic notions of this sort?

But Odile wasn't talking in quotation marks. "Muktananda is a perfected master of Siddha Yoga" she said. "By means of certain spiritual practices, he has reached a state of complete mental self-possession, therefore he doesn't have needs or wants. He lives in a state of bliss all the time."

I winced at the mention of bliss, with its overtones of empty-headedness and a silly grin. In fact, I wasn't truly listening to what Odile was saying. At the time I had a firmly established view of human nature. Fear was salutary, the gray ghosts of the night were my intimate friends. In a recent poem I had written: "I will praise the fear of death/ Which is the basalt of dark foundations."

Yet, in the end, I had become cozy with the "abyss." I felt that because of it, I had been guaranteed a sort of spiritual integrity. It would be impossible for me to "sell out," I thought, as long as I was so unhappy.

"Why don't you come and meet him?" she said.

There was no reason not to go. I lived only twenty blocks from the ashram and had plenty of time. Besides, in her quiet way, Odile was being insistent.

"He's quite an unusual man," she said, as if trying the words out. "In India they call him a saint, but I think of him more as a warrior."

Muktananda's temporary ashram in New York was a lovely red brick building, formerly a school, on 91st Street near Riverside Park where, I was told, he took long walks every day before dawn. Consistent with his aura, he had never been mugged. (Later I would have dreams of violent young men running up in the darkness to throw themselves at his feet. He would bend over and thump them on the back, or walk by, playfully raising his eyebrows.)

I was ushered into a medium-sized room with large windows, where a number of people were already waiting. Curls of burning incense, a foam rubber throne draped with a vivid cloth, a wand of peacock feathers upright beside it; people sitting expertly in the lotus posture, or leaning against a wall, or gathering their legs about them as best they could: the atmosphere was low-key, yet vaguely expectant. On the wall hung several huge photographs of a dark-skinned man in a loincloth, his body oddly smooth and glowing, his face expressing a combination of sleepiness and alert attention.

Muktananda never saw anyone privately, Odile said. At eleven every morning, he gave audiences to fifteen or twenty people. Even the disciples who traveled with him had to come to the eleven o'clock audience if they wanted to talk to him. That seemed as odd as anything else she had said. What sort of relationship could one have with a man one saw only in public, for a few minutes at a time? Surely he had some sort of private life. Odile assured me that he hadn't. He was always either alone or on display, you might say. Everyone around him existed on the same level, perpetually confronted with the mystery of his distance and impenetrability. Yet it was a mystery full of playfulness and spontaneity. He could be very funny, she said.

I didn't see the door open. He was simply there, quite suddenly. He walked across the room and took his place on the throne with a series of quick but fluid movements. Odile had warned me that he wouldn't seem very holy, and she was right. He wore an orange ski cap, dark glasses, and a gaudy robe. On the whole, he bore a slight resemblance to Dizzy Gillespie. His face, hugged by a short curly beard, had a kind of feathery alertness, as he settled onto his seat, checked a clock, tapped a microphone to see if it worked, looked for a pile of orange cards on a side table, and sprinkled perfume on the wand of sumptuous peacock feathers, occasionally darting glances around the room.

I had been startled when several people touched their foreheads to the ground as he came in, but I didn't really pay much attention, mainly because I didn't feel concerned. I wasn't there even from curiosity, I reminded myself, but simply as a gesture of friendship to Odile.

I noticed that the man in the enlarged photograph on the wall was not Muktananda, and asked Odile about it. She said that the photographs were of Muktananda's guru, Nityananda: a large-bodied, imposing figure, naked except for a loincloth. He seemed very different from the loudly dressed man fidgeting on his throne at the front of the room. There was a dark, almost demonic quality in the photograph, and a stillness which seemed to inhere in the figure itself.

Three Journeys

Muktananda communicated through an interpreter, a lively young man dressed in orange robes, who sat cross-legged on the floor at his feet. The interpreter called the name of each visitor to come up and be introduced. Not much seemed to go into an introduction. You got to say your name and what you did, while Muktananda tilted his head graciously and smiled. His smile was crisp and restrained, yet benevolent in its way. However theatrical his clothes might be, Muktananda's face did not indulge in flourishes. My turn came early in the hour. I went up and, observing what appeared to be a practice, got on my knees while the introduction was made. Odile, to whom I had given copies of some books I had written, dumped them on the floor in front of Muktananda, who picked them up and looked at them while the titles were translated. He asked if the word "emptiness" in one title had anything to do with the Buddhist "void." I answered that I had never thought about it. Did I want to ask him anything? That was the farthest thing from my mind. I said no, and the introduction was over.

More people were introduced. For the most part they were younger and had been involved in the Eastern scene in one way or another. The sorts of questions they asked rubbed me the wrong way; they seemed full of personal melodrama. "Sometimes I feel within me. . . ." "I know in my heart. . . ." "My inner awareness. . . ." "My cosmic feelings. . . ." I moved over to get a better look at Muktananda. For all of his sudden changes of expression, I began to sense a remoteness in his face, an immobility not unlike the face in the photograph. A young woman was speaking to him. She had lived for several years at an ashram in Pondicherry founded by Shree Aurobindo. She gave Muktananda a drawing she had done, and in a tremulous voice said she had a question to ask him. I found myself paying attention suddenly, not so much to what she was saying as to the note of vulnerability in her voice. When she meditated, she said, the experience of silvery light was very intense, but then nightmarish forms came between her and the light, and she was frightened. Her voice became increasingly tenuous as she talked, and then it broke. I could tell she was crying. When she lifted up a hand, as if to describe the nightmare, it began to shake.

Suddenly I was shaking too. I felt as if I were rooted to the floor, yet trembling with intense feeling. I had to make an effort not to cry, but it wasn't simply grief, for my body had become buoyant and warm. I stared at the woman's hand sketching a movement in the air. It was pale, delicate. Even after the hand was tucked away in her lap, and Muktananda's voice had begun to speak, I went on staring while the forms and colors of the room glided before my eyes like paper cutouts. The words "afloat in tears," repeated themselves over and over in my mind; an idea seized hold of me: all of us did our best against suffering and useless pain. Those nightmarish forms the woman had talked about were the element of my life, and everyone's life. All of us sitting in this room were on the point of crying out, for we existed far from the light.

As I stared at Muktananda's quick movements, I was aware that my mouth was hanging open, yet I couldn't seem to close it. For some reason I wasn't frightened; I was even pleased, though I couldn't say why. My jaws felt like hinged gates into a cave full of tears. Muktananda had done this, but what had he done, and how? We hadn't talked very much, and he had hardly looked at me. He was not especially charismatic: no great gestures or piercing glances. He moved around a lot on his throne, picked his nose, and played with his fingers. I became fascinated by his large stomach, which seemed out of place on a man of such slight build. And all the while I held my tears in by an effort of subtle attention. The tears seeped onto my face anyway, a few at a time.

Later in the hour I managed to stand up and indicate that I had a question after all. I marveled that my limbs still functioned, as I made my way to the front of the room. The atmosphere was dreamlike and filmy, and I felt strangely dissolved in it.

"My question is the same one you asked me earlier. What is the connection between the experience of inner emptiness, the frightening feeling that at some level of my existence I'm nobody, that my identity has collapsed and, deep down, no one's there; what is the connection between this feeling and the Buddhist void?"

"They are the same," he answered immediately, "but in the Buddhist void, there is no fear."

He said this offhandedly, and I felt somewhat rebuffed, yet I was moved, too. For until this morning I had believed that my insufferable anxieties belonged to the fabric of existence; that, in some way, they were a good thing. Now, sitting on a wood floor looking at a dark-skinned Indian with a large belly and an orange ski cap, I felt that a piercing light had been driven into my gloom: the sickness can be cured, it has been cured; you are already free. I was experiencing the delirium of release.

It was very much like a delirium. My head had become increasingly large and feverish. My thoughts floated in a syrupy atmosphere, over which my face seemed to be fitted like a distant mask. Yet the thoughts, surging like sea creatures, were strong and sharp. They were not my thoughts at all, but residents of the fluidity in my mind. I was a fisherman, a swimmer. Words like "happy" or "unhappy" meant nothing: something was breaking open; something was bursting like a pod; something was spilling out.

Toward the end of the hour, a nervous redheaded man was introduced. He stood in front of Muktananda, and began talking in a voice full of choked arrogance: "You people talk about bliss and liberation, but you ought to know that you're a tiny minority, a mere fraction. Most people don't see things your way. They suffer, and they hate. They work, and they feel frustrated. That's reality. What would Anatole France say about you, I wonder."

His body stiffened while he talked, and his shoulders hunched up defiantly. Every once in a while he squeezed a laugh from his throat, which resembled a cackle. He tried to get a cigarette to his mouth, but his hand was shaking too violently.

"What right do you have to talk about your bliss to people who are suffering? This is evil. Anyway, you can't prove it. How do I know you're not lying? You claim you're not afraid to die."

The Bright Yellow Circus

He cackled again. Then, as if forcing himself to speak: "Listen, I'm terrified of dying. What would you do if I pointed a gun at you right now; if I pulled a gun out of my pocket and pointed it at you?"

There was an undertone of violence in his voice that made him seem almost crazed. But neither Muktananda nor his interpreter seemed at all nervous. When the talk about the gun occurred, Muktananda answered: "My love would still be coming toward you while you pulled the trigger."

I remember thinking: this is preposterous, no one can say such a thing and mean it. At the same time, a thought filled my mind, as with a bright vaporous whisper: yes, it's possible; such a response is possible!

The redheaded man seemed to collapse. He threw his head back and laughed, almost shyly. All at once he was hugging himself and turning his body from side to side like a little boy. Everyone was laughing gently, and the man now seemed merely vulnerable, lonely. His fingers were stained with nicotine, and they still shook a little. With his talk of Anatole France, he reminded me of an uncle of mine: a nervous, frustrated intellectual. Then he reminded me of myself, or an aspect of myself: a frail, wiry individual who couldn't afford to be truly generous, who needed all his energy simply to stand still.

Muktananda glanced at a pop art alarm clock on the table beside him. He said something in Hindi to his interpreter, and stood up, glancing around. Everyone bowed, and he walked briskly out of the room. His way of walking was strange, yet marvelous, too. He leaned backward a little and swung his arms in a long outward arc. This made his round soft stomach especially prominent, as if he were being guided by it.

When I mentioned this to Odile one day, she laughed and said that it wasn't really a potbelly, but a result of breath retention. Most of the great Siddhas had round bellies. So did the Buddha. (Muktananda's stomach was to become, for me, an emblem of his compassion and power; a physical embodiment of the inward space he vanished into when, under his dark glasses, in the midst of a talk or a chant, he closed his eyes for a moment, and I would say to myself: he has vanished into his belly, he has disappeared into the source.)

After Muktananda left the room, we were invited downstairs and served lunch in what must have been an auditorium in the building's school days. I sat on the floor and thought the food into my mouth. The people waiting for lunch chanted in a language I'd never heard before. Their chant was rhythmic and full-bodied, not at all like a church song or a religious hymn. It struck me that these people were having a good time. At the front of the room, a group played rhythmic accompaniment on a drum, a harmonium, and a tall twangy instrument. The music fed the intensity which pulsed gently, making my body feel roomy and full. I had no idea what was happening, yet I still wasn't afraid or even curious. I was simply absorbed, as if I were dancing or singing with abandon. Yet I was sitting completely still. I might even have seemed sad to an observer.

I remember the afternoon in bits: trying to talk to a disciple, but not able to say more than a few words before tears overwhelmed me; Muktananda's appearance a while later, wandering across the auditorium to his seat near the stage. I had become so absorbed that it was a while before I noticed and hurried to sit near him, as did everyone else. Somehow he looked blacker, more solid. My eyes began to stare as they had that morning, tunneling deeply into themselves. Everything about him was so vivid: his wiry black beard and the moods flitting across his face. I felt myself dissolving again into the billows which broke warmly, silently in my mind.

Later I remembered what his disciple had said to me during lunch: "It looks like you've got it."

What had I got, I asked.

"*Shaktipat*, a dose of Baba's *shakti*, his energy. That's what you're feeling now. Baba says that all of existence is a play of *shakti*, but that our personal *shakti* is dormant, as it is in external objects. Being intensely aware of objects is equivalent to awakening the *shakti* in them. That is what Baba does. He activates the dormant energy in us. It's like a lamp being used to light another lamp."

The explanation didn't really make any sense to me. Merely to follow it as an explanation of something that actually happened, like the law of gravity, required a wrenching of my mental habits which was quite beyond me. Nonetheless, I was dumbfounded. Apparently other people had had this experience often enough to give it a name: *shaktipat*. This thought alone was full of wonder for me. So I was not simply flipping out, according to some personal law, the end result of which was crazy, if pleasureful; I was experiencing something real, something with a name.

This reassurance led me to glimpse a mental trap I had lived in all my life. An experience became real for me only when I identified and shared it by giving it a name. All my life I had read books, studied them, eventually written them, and the enveloping purpose of my reading and writing had been an endless anxious quest for words capable of communicating to me the reality of my own feelings. Adam named the animals according to a technique which had, apparently, been lost to me, for his animals stayed named, while mine sank back again instantly, so that nothing was ever gained.

Even on that first day, walking home along Broadway in a state between dreamy relaxation and pure aerial energy, I sensed that my system had been overthrown because what I was experiencing was simply irrefutable. This upheaval didn't need me to "prove" its reality. On the contrary, it was proving my reality, just as fear or erotic excitement are tremendous proofs of one's reality.

The energy fusing from every part of my body sufficed to itself. It wasn't so much beyond words as it was alongside them, in some other realm. I liked that idea. In a confused way, it increased my feeling of self-respect. My secret sprayed itself through my body, and I reveled in the misunderstanding which made me seem perfectly ordinary to anyone who saw me go by. Yet I had become different. My life was no longer subject to the universal law of suffering. I had escaped by some miracle which I connected to a dark-skinned man in a ski cap, whose precise features, movements, and voice already seemed a little blurry. For that very moment—filled with the amazing clarity of store windows, of faces streaming toward me like separate pieces of a single awareness—was so much more real than any place I could be coming from or going toward. I was aware of the strangeness, and possibly the ridiculousness, of these thoughts which insisted on coming into my mind. I could not help but take my feelings seriously, but it seemed best not to take these slightly mad ideas seriously. There would be plenty of time to see later on.

I was more a witness than a participant in the events which filled the rest of the day. The sound of my voice, the old-fashioned furniture in my living room, my wife's puzzled expression as my voice tried, and failed, to make any sense of the day's experience: all of this took place in a strangely tender realm, orbiting far away from the luminous bead at the center of my mind. I had the impression of being the hub of a world of events which "happened." This world included my bodily sensations, the ridges on my fingertips, the feeling of my wife's body as I hugged her with more affection than I had ever felt before; my wish that she know exactly what was happening to me; the black velvet sensation of making love later that evening; the mattress under my back and the lofty shadow of the ceiling as I lay waiting for sleep, and stared, not wondering or thinking but simply staring, at the only thing in the world: a dark, slightly peeled ceiling in the bedroom I had slept in for many years, which had been a theater for love, for my favorite anxieties, and my reluctant, nervous waking up on all those mornings which had been circles on the calendar of my **life**.

Earlier in the evening a **friend had** come by to talk about some poems he had written. I told him that my mind was a bit foggy, and I wasn't sure I'd have much to say. I mentioned my encounter with Muktananda that morning and the experience I'd had and was still having. My friend looked at me carefully.

The Bright Yellow Circus

His own life had taught him a great deal about the tricks the mind plays. He was not a man who had much faith in sanity, so it was with real concern that he wondered if I shouldn't talk to a psychiatrist. I found it odd that I hadn't thought of it myself, but I hadn't, and his saying it gave me a jolt, because I was filled suddenly with a conviction that, come what may, what was happening to me was absolutely sane, saner probably than anything I had ever experienced.

As we sat and read my friend's poems together, I had an experience of lucidity which was momentarily frightening. Something, not me, was reading those poems. Something was seeing what they meant as if the meaning were a map drawn to perfect scale. I saw the locations where the meaning faltered, and the poem's energy was deflected. I saw where the words strained for a farther, more secret meaning. It was as if my mind had gone into overdrive, and I sat back, watching its luminous swoop, hearing my voice suggest this, insist on that. I was generally fairly good at this sort of thing, but not this good. I felt as if I were inside those poems, and inside my friend's life. There seemed to be no limit to the mental penetration of this "something" for which "I" was apparently a lens focusing it into the world, making it visible and audible.

The same "something" is writing these words, months later, beside a frosty lace of moisture on the plate glass windows of the café where I come on afternoons, feeling as if a thumb were pressed between my eyes: a slight, confident pressure which signals the onset, as of a gust of inner wind, of the *shakti*, the playful coming and going of the meditative state which rarely leaves me entirely these days. It has snowed all day, and the streets have a white, crumpled look. Cars, trucks, and people move by in careful slow motion. The frost patterns on the window are incredibly suggestive. They are silvery, and they glow in the diffused sunlight. Looking at them, I feel as if my eyes were closed. It is a feeling I often have when I write about my experience of Baba. At such times, I have no idea what will happen next. My experience of time expands and becomes smooth. Going nowhere becomes, momentarily, a luxury. Everything has the glow of water beaded up on a windowpane: my hands relaxed on the table; the faces moving at people-height along the sidewalk; the indoor plants suspended near the window like clusters of green snowflakes.

Odile had suggested that I come to the ashram in the morning, to see what the daily activities were like. There was an early meditation at 6:00 A.M. I woke at five, and walked the twenty blocks. In the early morning darkness, West End Avenue resembled a vast indoors I strode through, to the door on 91st Street through which I stepped out, not in, to a candlelit vestibule. I left my shoes and coat, and went up a narrow flight of steps to the auditorium. The dense warmth of the room was full of breathing. It was completely dark, except for a relaxed dangle of candlelight along the walls under several large photographs of Muktananda's guru, Nityananda. I stood near the back, until I began to make out dim bundles scattered about the floor. I remember thinking they looked like headstones. They were people sitting in meditation. I didn't know how one went about meditating, but I sat on the floor with my legs crossed and closed my eyes.

It was probably the longest hour of my life. I thought about my aching back, knees, and ankles. I thought about my thoughts which wouldn't stop, or even slow down: perfectly ordinary thoughts squirming about like fish in a pond. The harder I tried to catch them, the more they squirmed and multiplied. I felt increasingly helpless, a bystander in my own mind. It was like a bad home movie, composed of endless reels of "shots" I never wanted taken in the first place. None of them had anything to do with my feelings of yesterday or this morning; none with Baba, or *shaktipat*, or the curls of incense unfurling in the dark room, except if I made an effort of will, which seemed out of place, for the willed images felt self-defeating and gradually depressed me. If there was ever a lesson in the futility of mental processes, I was receiving it. When the lights came on, and the soft sound of chanting started up, I felt as if I had risen to the

The Bright Yellow Circus

surface of a pool of foul water. Already my discouragement was dissolving. I heard myself laughing softly as I watched people form a breakfast line, chanting as they had before lunch the previous day. As I sat in my place, large-headed and giggling, it became clear that I had been meditating after all. The taste of trivia lingering in my mind was strangely happy.

How odd, I thought. If *shakti* is energy, then my anxiety was *shakti* too, and it was normal for it to turn into laughter, for it had been laughter all along, or rather, it had been neither anxiety nor laughter, but simply energy focused first through one emotional lens and then another. My life seemed totally open, as if I had been transported to a world which was governed by the law of total surprise. The only way to exist in this world was to start it up again with every glance. Existence threw itself into the air every minute, and came down in a new arrangement, damp with its birth water.

After breakfast there was a long session of chanting and prayers, in Sanskrit. It took many days before I could even follow the phonetic transliteration of the chants, and more days before I could join in with more than an occasional mumble. At first the chanting felt stagy to me, like some sort of transcendental circus. But gradually it dawned on me: apparently this is what people do around Muktananda. They chant, meditate, and stand in line for meals. When he comes into a room, they touch their foreheads to the floor; some throw themselves full length, and touch his throne with their hands after he's gone. They bring him little presents of fruit or flowers. They ask him to bless personal objects. They go up to him at every possible excuse to get touched by his peacock feathers. They ask him questions sometimes, not so much to get an answer as to have his attention focused in their direction for a while. Even his anger, full of shrill electric wavelengths, seems to thrill them. Clearly, from every point of view except the one I find myself in, this is absurd and silly, maybe even dangerous. It amounts to worshipping a human being, which is barbarous. In that case, why am I doing it? For I too bowed, got brushed with peacock feathers, and lost myself in the energy of chanting. I too found my eyes locked on Muktananda's face, on his wonderful brown hands and his supple legs. None of these activities made rational sense to me. Yet I kept remembering: while sitting in this man's presence, I was cured of a "nightmare" which I had come to accept as the main premise of my life. At this very moment, I am happy. In that case, why shouldn't I chant, bow, and beg for attention?

It was Pascal's bet, and I was making it with a great laugh: if Muktananda is what my experience tells me he is, then I've gained everything; if I'm deluded and it's all a fraud, then I'm back where I started, which is nowhere, a place I know all about. Besides, I had already tried the other solution, which consisted in being reasonable, filled with brittle dignity, and resistant to changes in habit.

To tell the truth, making the bet was easy, because everything I did set off tremors of fulfillment. When I repeated the mantra he had given me—*Om Namah Shivaya*—a molten feeling would come over me and I would find it hard to speak. Gusts of exuberance would start up at odd moments of the day. I thought about the Christian paradox enunciated by Tertullian almost 2,000 years ago: *credo quia absurdum est*, with its overtone of loneliness, struggle, and uncertain fulfillment. I felt as if a crack had opened in a wall I never knew was there, and I had slipped through it: on the other side of the crack, belief was a simple matter, like tasting or touching. It was a question not of faith, in the Christian sense, but of experience. I believe because I am experiencing it. I would encounter this same sentiment in a poem by the Hindu poet, Kabir: "If you have not lived through something, it is not true."

I read a talk by Muktananda in which he said that the sound specific to every activity, being, or event is yet another way the world has found to repeat the divine name. The sound of a running stream, the clanking of a bus, a man shouting from the sidewalk to someone at a window, a dog barking, my own talking, singing, or breathing: all of these were different "names" for the

The Bright Yellow Circus

energy which plays about us and in us, and is called, by some, God. There were moments when I could hear this underlying unity with the clarity of a bell ringing inside my ear. They were moments of complete happiness, as if, between one instant and another, a stillness had opened without any thoughts, hopes, or acts in it, but only the stillness itself, which made me vibrate like a musical instrument stroked by an invisible hand.

I made a decision during those first days which has governed all my uncertainties since then. Because I couldn't doubt my experience, I might as well believe everything, or at least not disbelieve too quickly. It was a tall order, because strange claims were made by Muktananda's devotees, and stranger stories were told by Muktananda himself about the power of his own guru, Nityananda, who had died some years before: Nityananda's spirit answering a frantic plea to frighten muggers away in a city he had never visited; his voice speaking in a dream to a kidnapped Indian student, giving him instructions on how to escape from his abductors. A French disciple told me she had gone to Nityananda's shrine in Ganeshpuri a year ago and explained to his statue that her sister had been trying to have a child for twelve years: "Nityananda, if you want me to make any spiritual progress, you've got to help my sister, because I feel responsible for her happiness, and couldn't possibly free myself from desires as long as she is miserable." Within two months her sister was pregnant, and has since had a child.

It was said that Muktananda could know what was going on in people's minds; that wishes made in his presence were fulfilled; that problems, however difficult, were resolved. Although Muktananda had no interest in supernormal powers, and disapproved of yogis who made a business of showing them off, he let it be known that such powers existed and that he had mastered their use. In an account of his guru's career, he wrote that Nityananda was such a powerful saint, that powers—*siddhis*—danced attendance on him, often without his knowledge. When money was needed to pay workers for some task, he would tell them to look under a stone in the forest, and the money would be there.

Three Journeys

It was hard for me to know how to feel about such stories. Should I simply believe them? Or should I consider them as vehicles for Muktananda's mental power, his *shakti*, and therefore true, but only in the sense that their apparent meaning made it possible for the hearer to be receptive to the perfected energy of a Siddha master? Did they have an esoteric dimension which I would grasp later on? Or were they challenges which Muktananda threw out for his disciples to wrestle with, the energy of the struggle being useful, for it forced the disciple to define his relationship to his guru, and therefore to improve his self-understanding? Perhaps all of these answers were needed to account for the stories. Yet I find, as time has passed, that believing or disbelieving has gradually lost importance for me. Whatever sort of "truth" the stories contain, I find it hard to dwell on them when the most important of all *siddhis* is demonstrated every day in the experience of my own mind awakening to the "sweetness" of those mysterious bees Muktananda talked about one morning when I had offered him a jar of honey: "They are so tiny, a breath of wind could blow them away like specks of dust. But the honey they make is limpid, clear as water, and it is tastier than any other honey. Such tiny, tiny bees, and they blow away so easily."

A few weeks after I met Muktananda, I had to leave New York for a week to join my wife, who had gone to North Carolina to exhibit jewelry she makes. This meant giving up the supports of my experience: meditating, chanting every morning at the ashram, and sitting in Muktananda's presence. The thought of doing anything that might cause me to lapse into my previous state of existence was frightening. At the same time, it was clear that I would have to deal with my fear sooner or later, because Muktananda was planning to leave New York in a few months.

The morning before my departure I went up to him and told him I would be away for a week. He said: "Come back soon. Meanwhile I'll be thinking about you." The rest of the day I felt feverish and my head was unpleasantly heavy. I hung around the ashram for a while, then went home to pack. I didn't feel desperate or anxious, simply absorbed in my discomfort. That

The Bright Yellow Circus

night, I hardly slept. But when I woke up, around 5:00 A.M., I found myself exploding with energy. Every cell in my body seemed to crackle with exuberance and laughter. I drove eleven hours without a stop, singing the few Sanskrit chants I had memorized at the top of my lungs the whole way: absorbed in myself, in the landscape flashing by, in the glint of sunlight and the colored eggs of cars sliding past.

That night I woke up after only a few hours sleep, feeling terribly anxious. It is a form of insomnia I know well: I wake up in the middle of the night filled with painful thoughts and lie in bed, increasingly oppressed, until finally I get up, put the light on, and start my daytime self functioning—reintroducing the tiny sanities of purpose, projects, ideas, until the anxiety, like a dark shallow pool, soaks away.

I lay there, stunned by the return of this catspaw of the old nightmare. My eyes were still closed, but I knew I would have to open them and begin the familiar ritual of self-distraction. Suddenly a bright yellow light filled my eyes. The light resembled an enormous surface, covered with the baroque line drawing of a circus: trapezes, animals, trainers, clowns. I heard my own voice saying: "It's all a play of the divine energy." The voice meant that my anxiety too was a play of the divine energy, and I understood this instantly. Then the scene dissolved. But the anxiety was gone too. I felt full and still, and fell asleep again quickly.

As months pass, it has become clear that this was one of the extraordinary experiences of my life: the sign of a mental reality I had never even approached; a talisman as solid and palpable as any outward miracle could be, letting me know that a mental realm existed for me to reach, that an experience of knowledge was there for me to grasp, if I pursued the quest which Muktananda had made possible. When I try, I can still see the yellow glow with a circus scrawled across it, and on occasion I have seen other things: a sultry colored jungle; the looming eyeless statue of Shiva; a blue mosaic floor extending to the horizon. I have come to think of them as Baba's gifts, or rather, gifts he has enabled me to give myself through the practice of a particular form of love, mingling devotion to the guru, and total acceptance of myself and of all experience: the practice of meditation.

Sitting quietly in my study in the early morning darkness, my legs folded under me, the air pungent with incense, a lit candle providing a palpable image of the inner light, and exposing a photograph of Muktananda, I close my eyes for meditation and bow mentally in all six directions. Then I move my mental eye slowly from my toes to my head, imagining the presence of God in each part of my body. I have invented this ritual to create a border of concentration for the blanket of awareness which I pull close to me. I repeat slowly the mantra Muktananda has given me, as if each syllable had a taste. I gaze at the syllables themselves standing in my mind: *Om Namah Shivaya.*

It is like sitting in the most familiar of all rooms. Gradually the walls of the room become transparent, revealing a boundless space streaked with colors moving gently past. Sometimes I feel a pressure between my eyes, as if something were trying to get out, like a folded-up chick pressing against its shell. Sometimes the pressure grasps my cheekbones and temples, too.

Now the room is like a jar hanging in space. Now the jar is my mind, my very Self. Its limits no longer shut out what is beyond them. They aren't quite limits anymore. Maybe I'm not in a jar at all, but swimming directly in the vastness. For a minute it seems that way, and the impression is blissful, then frightening, and fright recreates the jar.

At first, meditation disturbed me. It was an experience of persistent helplessness. Even when my days were electrified by Baba's presence, I found the aloneness of meditation oppressive, because all the familiar attributes of my identity were left out. My quick-mindedness, my various facial expressions, the little games of humility I could not help playing around Baba: none of these, nor any of the other mannerisms of my existence, played a part in meditation. I understood that meditation was meant to provide leverage for psychic change; that it was a center of experience around which everything else was orchestrated. But this was hard for me to accept, precisely because everything else was so extraordinary, while meditation remained empty and anxious. Yet even during those first weeks there were instants of absorption followed by flickers of blue light. Sometimes the light lingered, and I would feel a sort of inner release, reminding me that I ought to persist, despite my discouragement. The changes came gradually. Little by little I realized that I was actually learning how to meditate, though it was not clear how I had gone about learning. The sharpest turning points seemed to come each time Muktananda left New York for a while and later when he moved to Florida and I visited him there.

Each time he had said, "Leave-taking is not very important. You need only gaze inside yourself, and you will find me present."

At one point I told Muktananda about my experience of the yellow circus. I asked him if it had happened because he had actually been thinking about me, as he said he would before I left. He laughed and slapped his thigh. Glancing at me archly, he said: "Yes." And laughed again. Everyone else in the room laughed too. And it seemed to me that each person had heard what he needed to hear: yes, Baba can project his thoughts hundreds of miles and appear miraculously to his disciples in a moment of need; no, Baba obviously wasn't thinking about him, and now he's making a little joke for his and our edification. What I heard as if it had been spoken, was this: "I won't answer your question. My yes, combined with laughter, might also be no. Put another way, the mental reality of your vision was perhaps my thought, yet it came wholly from you, and was, therefore, your thought. It was mine and yours too. Mine, to the extent that you succeeded in forging a link between us: that link was the vision."

During the first weeks after I met Muktananda, my exaltation was so intense, so sustained, that I wondered, almost wistfully, whether I had stepped outside the human condition. I felt drawn to the faces I saw in the street. Faces were visible souls, and I was penetrated by the work which life had accomplished in them, as if an inner wind had dislocated the original elements, leaving a sort of wreckage: lips reshaped by tension, eyes set into dark smudges, cheeks hardened by anger or frustration. I seemed to see the traces of breakage everywhere. Yet I experienced even this as part of an immense, almost audible blossoming which was going on all around me, and in me. The breakage too was part of a flow, as if the world had become the bed of a river. The faces with their lives written into them in the signatures of wrinkles and discoloration, seemed bottomless to me; they offered no resistance to my seeing, which sank into them. At the same time, the seeing set me apart, not exactly lonely but ready to be lonely.

I had always believed that people were linked to each other by means of small fractures. Through all the broken places, an inner substance welled, mingling with the inner substance of others. Society, companionship, even love, was a miracle resulting from pain and, in a sense, failure. Yet here I was feeling healed and buoyant. At such moments I felt as if my being had been displaced into a different continuum of existence; as if, in some important sense, I was no longer here.

Yet, as far as my friends were concerned, nothing particularly noticeable had happened, except that I had, on occasion, a new and rather bizarre subject of conversation. When, now and then, I saw myself with their eyes, I felt like the character in the cartoon who has climbed a ladder; after a while he looks down and notices the ladder is gone, he is standing on nothing. As the days passed I discovered that I had begun waiting to fall.

Two occasions stand out in my mind from these early days. One is my first, and only, "private" encounter with Muktananda. One morning I arrived at the ashram feeling depressed and frightened. I had gotten out of bed, and there had been nothing. As if a light had gone out. I had been waiting for the fall, and now, apparently, it had come. This dark-rimmed, overly complicated person was me as I had forgotten myself to be: slightly scuffed, anxious, and lonely. My meditation that morning was boring. My tongue felt like a piece of rubber. My throat was sore. Muktananda was present during chanting, and I kept look-

The Bright Yellow Circus

ing at him, thinking: what could he know about suffering? Why does he parade his bliss around in a world full of broken people? There was something callous about it. Yet I wasn't really angry. Maybe I didn't have enough energy to be angry. After the chanting, one of Muktananda's aides asked me to come upstairs to look at a text they were writing. I followed him, feeling the absurdity of my dejection and wanting to get out of there quickly. Going up the stairs, I noticed Muktananda coming out of his room. He stood in the open door, his hands behind his back, and looked up and down the hall. Before I could think, I had gone over to him, dragging the aide, Yogananda, with me to translate.

"Baba, I'm feeling so depressed. What's wrong with me?"

He darted a look at me, and gave me a friendly tap on the chest, laughing: "What you're feeling isn't really depression, it's indifference."

He laughed again, and then pounded me on the chest, repeating: "Not depression at all, indifference. Yes, very good."

The last words were said in English, his comical accent making a joke out of them. Then he turned, laughing to himself, and went on down the hall.

I continued upstairs and did what Yogananda had asked, feeling jittery and impatient to leave. As I clumped downstairs and put my shoes and coat on, I could still feel the impact of Baba's hand on my chest. His high-pitched laughter kept turning in my mind. With no transition, I was running up the street, wanting to chant at the top of my lungs. An exalted feeling brimmed through every pore of my body. All day I wondered—had it been the words, the laughter, the touch, or something else? The connection between depression and indifference was interesting, to be sure, but how could that provoke the intense happiness which flared through me and diminished only gradually into normal lightheartedness?

On another occasion, I was having dinner with a friend who is in the habit of asking momentous questions over cocktails. In the past, always when least expected, we had talked about the nature of love, about the erotic attraction of extremely young girls, about Christ, about the meaning of the word "nature," etc.

"Paul, why is the imagination such an important part of mental life?" he asked, puffing on his pipe, on his face a painfully earnest expression combined with a suggestion of slyness. "Why do we all assume it's such a valuable faculty to have?"

As the question came, I felt myself grow excited by the power of an idea which imprinted itself in my mind with exhilarating clarity.

"You can think of the imagination as a form of play," I said. "Essentially, it is a way of playing at self-destruction. Through the imagination, we master the actual prospect of destruction by making it happen over and over again, like the game Freud once saw a child playing. The child got great pleasure out of making something disappear and reappear endlessly. It turned out the child was experimenting with loss, in particular the loss of his mother who, naturally enough, went away sometimes and then came back. In the case of the imagination, what is 'destroyed' is a portion of our identity which normally is kept in place at a great expense of mental energy. It is the identity of willpower, of making plans, of calculating and foreseeing, of remembering; the identity which manages social behavior and knows about causes and effects. It takes a great deal of work to keep this identity, which we normally think of as our 'self,' intact. We have a feeling it might collapse if we forgot to think about it, so we devote a large portion of our life-energy to insuring its continuity. As a result of this exhausting labor, we find we sometimes need a holiday; we need a seventh day, when the Lord rested. The imagination is our seventh day. We play at giving up. We let go, the house falls down, and infinite quantities of space pour in with a feeling of plenty, of wealth. For a minute self-destruction feels like the only way to really live. The space pouring in takes the form of images, of impulses, of spontaneous thoughts, memories, inner fables. That's the weather we built the house to be sheltered from. And for a split second, we learn two things: it feels good to be destroyed; it feels equally good to notice the house miraculously intact, not destroyed at all, but made somehow receptive to the weather."

My idea had begun as a kind of ironic paradox, appropriate to a mock-serious conversation over tall drinks, but the paradox had taken over, as if I were being spoken through by a self I had never met. When I stopped, my friend cleared his throat nervously, and took a long puff on his pipe:

"Have you been thinking these matters over for some time?" he asked hesitantly. His question underscored my feeling of bewilderment. No, I hadn't; I had, in fact, learned them that very minute, just as I had learned "it's all a play of the divine energy," by listening to my own voice giving its opinion. Only months later did I realize that an explanation had been offered to me that evening, concerning the nature of meditation and "bliss": both resulted from a release of psychic energy which was no longer trapped in the labor of self-defense. Imagination, sleep, the delight of full self-realization, were conditions of total vulnerability which was achieved by alchemizing all danger into an experience of energy: not a threat to one's mental integrity but a part of that integrity. It was a question of learning to be part of the weather. Sitting in meditation, one fed one's experiences, one's thoughts, all the dirty wash of a day's complicated living, into a great laundering device: what came out was pure being, yours and the world's undifferentiated.

Several times since meeting Muktananda, I have had the experience of being spoken through by an idea which expressed itself more to me than to the person I was talking to, as if an inner teacher were addressing me with my own voice.

Only after my initial experience had begun to subside did I fasten my attention wholly on Muktananda. Before that, I had been too involved in watching myself live. To feel was about all I could manage, while the mental furniture moved itself around, and an earthquake rippled all the old corners of my being: an earthquake of previously unknown emotions—of "ice cream," to use Baba's expression; a storm of "sugar" and "honey," in the words of the great Sufi poet, Rumi.

At the center of the change was Baba, seated on his foam rubber throne, his head darting from side to side, singing verses of poetry from Indian scripture or from the work of the great Siddha poet, Kabir. He touched long lines of visitors with his perfumed peacock wand, gave out candy, and transformed the room into an echo chamber with his high-pitched angers which seemed to fasten mostly on small matters: a broken microphone, someone slipping into meditation during chanting, or coming late to a meal. The angers flared up and calmed almost instantly. And they were never predictable.

Every morning I sat as close to him as I could, noticing when his eyes fluttered closed now and then, watching his lips move during chanting, and following the pattern of his long graceful fingers on the tambourine he often played. Very few words passed between us. In fact, after the first weeks, he seemed not to notice my existence any more, except for an occasional glance in my direction from under his tinted glasses. I never knew about those glances. They seemed to be filled with compressed rage. But were they even glances at all, and if so, were they at me? And was it rage, or could it be something else: a signal, the equivalent of a hand reaching into my mental apparatus to accomplish some subtle retuning?

The more he ignored me, the more he came to resemble something like a pure phenomenon to me: a sacred tree, or a fire in a vast cold place. Everything I did at the ashram became a way of organizing my presence in his vicinity. I stared at his face, listened to him speak, clustered at his feet along with the other devotees, like a flock of birds huddling around a fountain. His presence was so stark and dense, it seemed as if he were composed of another sort of flesh: more opaque, more magnetized; not flesh at all, but a shimmer of condensed energy, assuming this dark-skinned, uncharismatic, yet totally compelling form.

I started to think less about myself, and more and more about him. And as I did so, the nature of my meditations began to change. Meditation became a form of profound aloneness, and at the same time a conduit to Baba. In meditation I felt closer to him, sometimes, than in his actual presence, when I would often be sidetracked by annoying questions: "Why doesn't he pay any attention to me? What did I do to make him seem so reserved? Why can't I focus my attention on him without being distracted by so many self-serving thoughts? If he really can read my mind, then he's got to despise me a little right now." The futility of these thoughts was clear to me, yet Baba's presence would often stir them into furious motion.

I had to admit to myself that it was no easy matter to hang around Baba; that it was sometimes even a little unpleasant. The experience was humbling, because I could see how childish I was being, yet I couldn't help it. At the same time, Baba's example provided me with a new resource: it taught me to be a bemused spectator of these silly notions of mine, watching them flit about with the tolerance of a tender parent, for even they were an energy directed at Baba.

During the last weeks of Muktananda's stay in New York, I began to spend five or six hours a day at the ashram. The rest of my time was an interlude. Mostly I sat in my study, my thoughts moving in an unfocused way among the impressions which had accumulated during the day. They were satisfying afternoons. I did almost no reading. Even Muktananda's spiritual autobiography, *Chitshakti Vilas,* couldn't hold my attention for long. I played Gregorian chants and Indian ragas over and over again, and I waited. I don't think I was waiting for anything in particular. My usual circle of preoccupations had simply drifted an arm's length away; cradled in the space they had left, I watched them float around me with a lazy eye. The sensation was, perhaps, that of digesting a subtle food: my laziness enabled Baba's nourishment to penetrate the hidden parts of my awareness. It resembled an attenuated form of meditation.

I am surprised, now, to recall how much of my ordinary life was carried on quite actively during those days. Though I was on leave from my university, I wrote articles to make money, and continued work on a book. I had not become a hermit sitting cross-legged in a cave, but I was in a mental cave which never lost its solitariness, even in the midst of a dinner party or other social occasion.

Friends tended to assume that I spent hours in intimate conversation with Muktananda, that I found him to be an extremely wise person with whom I had discovered an unusual affinity. How else could one account for my embarrassingly intense devotion to the man? Indeed, how else? Yet they were wrong. He virtually never talked to me, and when he did, I rarely had anything to say in reply. We have never actually conversed. To tell the truth, on the rare occasions when he has talked to me, I have been so filled with excitement that I could hardly listen to what he was saying, and have had to reconstruct it later from the memory traces of his voice.

Clearly my bond to Muktananda was not a relationship in the ordinary sense. I didn't "know" him at all, and if he "knew" me, I had a sense—disquieting but also reassuring—that it was the way a candle might be said to "know" the objects its light falls on: with an intensity dependent only on the distance of the objects from the flame and not on their particular qualities. I couldn't "deserve" Baba's attention any more than I could deserve light. Yet his light shone without interruption. It was not shed on me by means of personal gestures, advice, or any of the forms of human giving and taking one normally thinks of. But its shining was there, so definite and precise I rarely had a chance to doubt it.

The more I tried to solve the elusiveness of Baba's character, the more I became intrigued by my failure. My occasional stabs at thinking have led me to the conclusion that Baba doesn't have a character in the way we normally mean. He has many characters: every mood and mannerism is exhaustive and wholly articulated. The reason why one's awareness focuses on him so intensely in his presence is that he is himself wholly present. None of his mental energy is reserved for holding onto a past or planning for a future; he is never saving, always spending. Every movement he makes, from beating on a tambourine to scratching himself, is so decisive it rivets our attention. The thought and the act coincide; the want and its fulfillment are the same. Another way to say this is that he has no wants, for nothing in him is incomplete. It is, from this point of view, an existence bathed in magic, a spontaneous existence. Maybe that is the fundamental *siddhi*—or supernormal power—which Siddha masters are said to possess: the magic of unwithholding presentness. When we meet Baba, we set about fitting our impressions into a unity, as we usually do when we meet a person. We call this unity the person's character, meaning by that the limits he creates for his life by remembering and wanting in habitual ways. One's character is one's history. But with Baba, our effort doesn't seem to work. The unity doesn't telegraph itself to us. When we insist, thinking we've made some error in our arithmetic, we end up looking with puzzlement at this weird activity of our mind which, momentarily, resembles a sort of dancing after the music has stopped playing.

I have come to wonder whether this tantalizing unrelationship is not at the heart of Baba's teaching. Its aim, perhaps, is to isolate our mental activity by prompting it to sprout vigorously, boisterously. Meeting Baba, the disciple throws his emotional resources into the accustomed labor of building a relationship, only to find, instead of a psychic network mingling with his own, a compassionate, impersonal light. He had never seen himself in such a light. The furious burgeoning of his thoughts reaches into nothing; the familiar dance suddenly seems strange, because the partner is not dancing, and, yes, there is no music.

Previously it had seemed only natural to wrestle with the world, because the world had answered with wrestling of its own: hate, desire, worry, anxiety had grappled with hate, desire, worry, anxiety. The disciple's emotions had lunged against opposite emotions, also lunging, and the resultant tangle was called a relationship. How else to behave? But now, a glimpse at a time, the lunging seems isolated, absurd. One doesn't know how to stop, but the slippery creature no longer seems quite so inevitable, quite so intimately tied to one's identity.

Three Journeys

In Baba's presence, for the first time, I began to perceive my double: the obstructive network of wishes which duplicated my existence and projected it, like a shadow, into the world ahead of me. I walked behind this double, always in its shelter. My double received the brunt of the human storm, not me; my double arranged for love and disappointment, for worry and trust. I, a step behind, received only the diminished ripple of these experiences. My double was constructed of mental busyness; it was glued together by obsessive thinking and propelled by anxiety. The commotion which my double specialized in making had a way of drawing attention to itself, especially my own attention, so that I tended to be overly involved in promoting my double's activities, as if I were, indeed, my double, and little else.

Of course, this was not wholly true, and the best evidence for another view of the matter was my writing, which I often reread with puzzlement. My poetry and prose often seemed to know things which I, clearly, didn't know. I had written about love and the nature of the interior self; even about meditation; even about the energy flowing in superpersonal billows through the psyche from the remote inward spaces out into the world, and from the world into the mental interior. But I had reversed the sign of this energy, for I had perceived it as a danger, an enemy. I had identified it as the crucial flaw of human and my own nature, which Christian morality had mythicized as original sin. I called this energy "anxiety," and saw the world as a dark analogue of its presence. "Dark" was my favorite word, "fear" was my oxygen, "insomnia" was my heroism. Yet, in a peculiar way, my writing had gotten it right, albeit reversed, and temporarily unintelligible.

For some time I had been unsettled by the poems I wrote, as by a self-portrait which I didn't quite recognize. Now, sitting near Baba's throne, watching his quick, dark face, I saw the weather-vane flapping in the psychic wind more clearly than ever before. I saw my double agitating in front of me: worry, ambition for a book I had just published, a mania for comparing myself with other disciples who seemed to meditate more intensely or chant

The Bright Yellow Circus

more robustly: these clouds kept shutting out my view of Baba, whom I wanted passionately to "see," and I began to grasp a new sort of truth: this agitated bundle of preoccupations was not all of me; on the contrary, it was always getting in my way. To be sure, before meeting Baba I hadn't minded very much, because I had learned to see myself and the world through its eye and with its values. But now I longed for my double to doze off for a while. I longed to be face to face with the steady warmth which Baba shone in my direction. And the more I longed, the more agitated my double became; for wanting, longing, and mulling over were its game.

Sitting in my disorganized lotus posture on the floor of the auditorium, I would find myself imprisoned by these thoughts. My throat would dry out. If I were chanting, I would find it hard to squeeze the words through my parched larynx. My body would ache from sitting in an unnatural position. The thoughts would become like a ball racing around and through me, absorbing me into its discomfort. Then, something would happen. As I lost myself in the anger and discomfort, the ball of thoughts would gradually speed up. Closing my eyes, I would actually see a dark sphere spinning, getting smaller and blacker, and then it would vanish. Sometimes this was accompanied by a trancelike feeling in which Baba's face would seem to float before my eyes. Sometimes, like a silent movie, there would be no emotional accompaniment. I imagined Baba packing the dark flimsy substance into an ever smaller ball, then throwing it away. When this happened, the nagging thoughts evaporated, and I would be left, as on a bright empty beach, with a feeling of happiness.

How Muktananda made these things happen, I don't know. But they happened. They were the live energy pouring through the unrelationship which Baba had created, or rather, had enabled me to create between us. Life around Baba became a sort of hide and seek. I rused and struggled with my double. I tried to leap over it, and into Baba's arms, and became lost in the murkiness of the encounter. Then, suddenly, Baba's mental hand would pack the double into a ball and throw it away. And I would glimpse another order of experience flowing in all the nooks of my awareness: an experience of pure and simple happiness. The happiness bound me to Muktananda. It strengthened my devotion to him, and my love functioned as a conductor for his love. The more I cleansed the circuits of my devotion, the more Baba's energy flooded through them. The more I felt drawn to him, the more I experienced myself in a new way. I began to understand what Baba had meant when he said that love was a gift you gave yourself. By showing my double to me, by teaching me how to pack it into a ball and throw it away, Baba was teaching me how to give myself this gift. Love for Baba, it turned out, was self-love of the most joyous sort.

Muktananda doesn't emphasize techniques and special disciplines. In terms of Yoga practice, he often says his "method" represents a shortcut: you simply sit back and let his energy irradiate you in the form of *shaktipat*. Once that has happened, it's out of your hands. The inner changes will unfold according to their own rhythm. You need merely be a spectator. Indeed, any role besides that of spectator would be a hindrance. Chanting, meditation, ceremonial activities, are merely ways of deepening your receptiveness, while the released energy, the *kundalini*, does its work.

It turns out, of course, to be harder than one could imagine simply to watch the changes happen. The role of such a spectator becomes, perhaps, the most difficult role in the world. Your entire identity seems to resist it. Until Baba's *shakti* touched me, I had formed the all-too-human habit of connecting my wishes to their fulfillments by means of willpower and positive acts. The notion that I was "doing" something, and the endlessly repeated experience of doing it, was crucial to my sense of personal integrity. Then Baba said something like, "Stop wasting your time. All this doing doesn't amount to much anyway. Do you really believe you are the one who's doing it? Well, here's one thing you can't do, so why not simply get out of the way."

So you try to get out of the way, and discover that you can't. You keep trying until, befuddled and a little desperate, you say,

"To hell with it." And that's what Baba had been driving at all along. Saying "To hell with it" with all your heart releases you. The spectator, sitting back, feels happy and free. Feeling happy and free, he begins to stir about again; he begins to jiggle in his seat, and make incorrigible little "doing" movements, and pretty soon it all begins again, until, after a while, tied into a knot, he says, "To hell with it." And it goes on, this learning and unlearning and relearning. That is Baba's easy path, that is his "shortcut." It is indeed as short as your stubbornness, which can be pretty long.

I last saw Muktananda in Florida, where I went to spend a week with him. The house he was staying at was located in one of those embalmed suburban neighborhoods in Coral Gables. The houses were bland, the people invisible. The tropical sunshine seemed to gain a chill as it poured between the rows of boxlike homes. In the morning and evening Muktananda took long walks, and his orange robes, from a distance, resembled a flame promenading in the deserted streets.

It was one of the more bizarre Florida holidays anyone has ever taken: no swimming and no sunshine, hardly any sleep, and hours of sitting on Baba's living room floor or in the hall where he gave audiences, my face vacant, my eyes sprung open, or drifting half-closed, or closing entirely to watch the inner movements of light.

I absorbed every detail of Baba's appearance: his graceful, busy toes; his smooth shins with little scars on them; his hands like musical instruments; his face; above all, his wonderful stomach swelling under the gauzy shirt he wore in the Florida air. I knew this to be a traditional form of meditation: internalizing the visible aspects of the guru, the disciple also internalizes the guru's perfected psychic life.

This form of meditation began of itself one afternoon while Baba was having his mail read to him. He was wearing less clothing than he had in New York, and his body seemed smooth and supple. Not at all the body of a man in his late sixties. His toes were playing with the carpet. They seemed like monkey's toes, each one with a separate, prehensile life. He had let his arm dangle over the side of his chair, and his hand stirred faintly in the air like some kind of brown plant. For some reason, all these details fascinated me as they never had before.

When I closed my eyes, a tremendous pressure squeezed my temples. My mental eye still saw Baba lounging on his chair. It began to move from his toes to his face like a lens, while at the same time melting my own body into each image. Those were my toes, my scarred shins, my belly like a soft sun, my pursed, busy lips, my fleshy eyelids closing for a few seconds under the dark glasses, my hand stirring in the air. Later, when Muktananda had gone back to his room, I lay down on the terrace and closed my eyes. My entire field of sight was filled with intensely glowing blue tiles extending as far as the horizon. A rush of cold energy went through my body, and I experienced a feeling of total wonder. I wanted to give myself up entirely to the experience, but I was afraid to, and after a short time the tiles began to dissolve into ordinary blue light, and then a faded white light. I opened my eyes.

Such visionary moments have been rare for me, but so vivid I have had only to think about them to recall them in complete detail. I think of them as signals from a neighboring psychic realm, mental cairns which I glimpse at crucial moments to keep from getting lost or confused on this path I have undertaken. Even my many doubts and occasional sense of foolishness seem to carry a message attached to them: if you're going to doubt, doubt for all your worth; if you're going to feel foolish, at least feel totally ridiculous; if you're going to be anxious, let it be the champion of all anxieties. Let your troubles burn themselves alive. Turn even them, especially them, into their underlying reality, which is energy.

On my last day in Florida, I was asked to give a talk about Muktananda and the experience of meditation. It seemed odd to be asked. What could I really claim to know about Baba, aside from my particular experience? And what did I even know about

The Bright Yellow Circus

that? I had not tried to find words for the changes I was going through. Words had gotten me into the "nightmare" in the first place, or so I often felt. For years my overly articulate nature had expelled experiences from my psyche a little too quickly, as if I had needed to turn my life into meanings before it could hug me too closely. The result was that my words contained emotions, understanding, even wisdom which came from me but were not mine. I was a hostage to my words: what I spoke, I didn't have; what belonged to my "style" did not belong to me. It had been something of a devil's bargain, and for a while I had been willing to keep the bargain: a feeling of personal emptiness in return for good language; a private conviction of failure in return for steady doses of attention from all the hearers I could induce to gather round. It reassured me that so little of my communication with Muktananda had been by means of words.

My talk was to be at a day-long session of chanting and meditation. I woke up well before dawn and sat in a corner of my room trying to meditate. A sharp ache radiated inward from my eyes and forehead; when I closed my eyes, breakers of gray light crashed and withdrew like snow falling on dark water.

We drove to the hall where the day's program was to take place, and I chose a place to sit near Baba's throne. As the day progressed all I could do was to chant, weep, and try to ignore the splintered feeling in my brain. Each time I looked at Baba, I winced: his movements were like blows on my forehead. I had been told that one got such headaches sometimes. They were *shakti* headaches, the result of all that inner furniture being moved about; to which I added a week's lack of sleep, the shock of leaving Baba, and the talk I was to give without knowing what to say.

Under my pain and panic a perverse thought fascinated me: if I were to make a fool of myself this afternoon, that too would be all right, because being foolish might be exactly what was required of me today. There was even a seductive logic to my thought: what in fact had I always been most afraid of? Being, or at least seeming, ridiculous. Therefore, if I fumbled for words and said foolish things this afternoon, if I even spoke disrespectfully of Baba, that would be part of Baba's cunning. It would be his "treatment" for my obsessive self-importance, and I would swallow the treatment bravely. I would find that it tasted weirdly good. Baba would glare at me in disgust, and that too would almost be a pleasure. I almost wanted it to happen. There is a kind of authenticity in making a fool of yourself, I mused.

My turn came in the early afternoon. I stood near the front of the room and looked at the faces crowded together before me. For days I had had the haunted feeling that, one by one, I had replaced the parts of my body with Baba's. Now, as I pursed my lips together and my eyes drifted closed, my face too became his. The warm, thick darkness I gazed at was the space of his awareness. The sharp throbbing ache behind my eyes was Baba's egg hatching.

"We generally agree that love represents a high value," I heard myself begin, "as in love thy neighbor, love thy parents, even love thy enemy." As I spoke I felt Baba's lucidity possess me once again. I had become a voice filled with his serene energy, as with a perfect weather. "Most of us would probably agree that love is our ideal emotion, and we would say it a little wistfully, because there have been only a few short times in our life when we have known, personally, the dislocating power of love. The rest of the time we find it necessary to preserve certain limits: to have affection, to like, to feel tenderness, to 'love' with civility and restraint, expecting the same civility and restraint in the 'love' that others feel for us. The other, more extreme kind of love isn't forgotten, but we idealize it by playing rose-colored lights over it and using only elevated language when we talk about it. We direct it toward idealized objects too: beautiful women who preserve us from the danger of love by not loving us back, parents who are safely dead, or Jesus Christ whom we visualize with the aid of highly stylized images painted by artists for whom He was already a stylized ideal 1,500 years old.

"On the whole we tend to feel that an ideal is somewhat unrealistic and therefore not wholly serious. Our reverence for it doesn't require that we change the way we live because the beloved ideal is hopelessly remote from our imperfect existences. The ideal is like a star, keeping us company from a distance. Because of the distance it is safe from our bad influence, and we are also safe from it. Love represents just such an ideal. We think of it as a heaven inhabited by beautiful forms and lovely words which we glimpse, and even approach, but only now and then.

"Yet this ideal, so far from the normal emotions of our lives that we have to select especially remote and radiant images to represent it, is marked by an almost forgotten trace of an entirely different nature. Even the most ordinary popular love songs may convey this trace, in the form of an undefined longing which can be overpowering, almost religious in its insinuating attraction. Even a pop tune has the power, sometimes, to make me feel like an exile wandering about in an empty world. In exile from what? From happiness, or from great sorrows; from whatever is the opposite of loneliness and vulnerability and politely limited emotions. We listen to the song and for a minute the ideal isn't rose-colored anymore. It sinks its teeth into us like something hungry that would break apart our lives if we let it. All the great legends of love and death mutter and turn over in our psyches, and we hurry to put them to rest, because delirious, total love has no place among our practical values. Our capacity to cope with it has never been truly developed. That is why the Sanskrit word, *bakti*, unwithholding love, has no real equivalent in our language. 'Devotion' is as close as we come to it, and that is pale and dignified compared to the total sweetness of *bakti*.

"Yet there actually was a time now forgotten when *bakti* was the ruling emotion in our lives. Torrents of *bakti* gushed from us in a wild unreflective spray. We were sluices of inner *bakti*, and the objects of our love were looming figures, possessors of power, mysterious, unpredictable. Their being was on such a vastly different level from our own that only the maddest expense of energy could scale their heights and form a relationship. Of course, this was when we were infants, and the objects of our infantile *bakti* were our parents. Never again have we loved as we loved in those days. Our lives pulsated with love. We cast it like a net far beyond ourselves, and caught in its mesh those great bulks which we grappled close and thrust within us. Yet there was never enough. We were always hungry, always loving. We could perceive the world only with love-eyes, which hurt strangely when the world resisted our need. Our *bakti* was so powerful, it hit at last upon a way to solve its hunger. It would transform those resisting, unpredictable objects, our parents, into interior presences. It would rebuild them, like miraculous sculptures, within our mind, where they would actually be ourselves. Our thoughts and wishes, even the way we moved our bodies and organized our feelings, would be acts of interior *bakti;* endlessly repeated *hommages* to our lovers.

"Around this time, however, we made a discovery. The forms we had erected in our minds and worshipped until they had become the very structure of the mind itself had been created all too faithfully to their outward originals. Their style and dignity had become the foundation for our own style and dignity, but their painful contradictions, their anxieties and hopelessness, all the normal flaws of adult human nature, had also become foundations for those very same traits in ourselves. Along with our parents, it turned out that we had swallowed a strange poison, which they too had swallowed, and their parents before them in the same way.

"Our sense of betrayal must have been extraordinarily painful, because we resolved never to surrender ourselves again to the internalizing power of *bakti*. We would preserve our reverence for the love which had taught us how to think and feel, but we would insure ourselves against its recurrence by turning it into an ideal.

"Now suppose that sometime in our lives, we were to meet a person who had confronted precisely the painful underside of his humanness, and had managed to eliminate it, as the body eliminates the waste part of even the most nourishing food. Having learned the nature of the 'poison' he had found an antidote causing it to pass gradually from his system. In many ways, he is still very much like the rest of us. He eats and sleeps, gets angry, is subject to sore throats and toothaches. His personality is rich in contradictions. He grows old as we do, and has his caprices. But precisely those self-defeating needs which our infantile *bakti* had smuggled into our characters have been eliminated. So that when we meet him, our elaborate defensiveness doesn't clash with the expected weaponry of an opponent. We defend ourselves according to custom, but there is no attack. This had never happened before. We test again, and again the mental counterthrust is missing. We are thrown off balance and topple forward, into his arms so to speak. Before we know it, we are in quite a muddle. Nothing is going according to plan. Our sensible restraints are flapping about at loose ends; on the one hand they don't seem to work, and on the other they don't seem needed. In fact, we are not exactly sure if our situation represents a dilemma or a liberation. All that we know is that everything seems changed. Our emotions are running riot. Our perceptions and our self-awareness pulsate with a feeling of plenitude which is so exotic we aren't sure we know how to cope with it. So much happiness almost seems threatening. We feel the deepest layers of our psyche tremble, as from an earthquake; or rather, tremble again, as they had in some not quite remembered past.

"I guess I'm describing my own experience of meeting Baba a few months ago. I'm describing the experience of possibility which burst in my life, and which many of you may have felt too. Since then, it has dawned on me that a lucky second chance has been offered to me. Having chosen once upon a time to keep my losses to a minimum; to renounce the vivifying love which forged my being, but also produced the inner flaws of my nature, making it necessary for me to live with anxiety and disappointments; having chosen to love only in civilized doses, and therefore to become a socialized, adult human being; now, I have discovered, an awesome and wholly unforeseen occasion has arisen for me to release the suppressed coils of *bakti*, to become in this respect as a little child again, casting the net of love and hauling within me yet another miraculous sculpture of another parent,

Baba, the father of my second birth. This new parent has extirpated the ancient poison from his nature. Therefore, by accepting him in my mind, I am injecting, or introjecting, his perfected existence, to counteract and expel the inner harm which had convinced me, long ago, to renounce the torrential wealth of *bakti*.

"Why do you think the apostles hung around Jesus all the time? Why did noblemen and peasants, during the Middle Ages, have such awe for hermits meditating in their caves all over Europe? It was because they understood the psychic perfection these holy men had achieved. They knew that by simply hanging around such men, they too might get their second chance.

"A Siddha master like Baba is, in the most profound sense, a teacher. But what he teaches is not a philosophy; it is not a system of knowledge or behavior, nor is it expressed principally in the form of wise sayings or poetry, though both are implicit in his teaching. He teaches only himself; he is himself the lesson. That is why his devotees call him Baba—Father—and feel such love for him that they are endlessly fascinated by his voice, his movements, every nuance of his presence. That is why we trust him. We have learned from experience that by trusting him we lift ourselves, degree by degree, toward the psychic balance he has achieved for himself. Trusting him we trust ourselves, and loving him we love ourselves."

During the weeks that followed, it occurred to me that the talk I had given in Florida represented a turning point in my experience. This was as far as words would take me. I had been lifted as by a gust of wisdom, which had set me down beyond the reach of eloquence. Words had carried me as far as the beginning. Now, in a far different sense, it was up to me to begin.

PS
3576
W4   Zweig, Paul.
Z52   Three journeys.